Surveying

Dr.G. Venkatesan

M. Aarthi

Published by

BONFRING®
Intellectual Integrity

Surveying

ISBN 978-93-89515-33-6

Authors

Dr.G. Venkatesan

M. Aarthi

Bonfring

309, 5th Street Extension, Gandhipuram,

Coimbatore-641 012,

Tamilnadu, India.

E-mail: info@bonfring.org

Website: www.bonfring.org

Preface

This book "Surveying" provides the fundamentals of various techniques of surveying. This book is based on the latest syllabus (R2017) of Anna University, Chennai, for the Third Semester Civil Engineering Students.

Surveying is not only applicable to constructions. It is also used in implementation of highway and Railway projects, Water supply Pipelines etc. This book provides an extensive idea on Surveying and its practical applications.

Unit I: Classification and basic principles of surveying, a details study on chain surveying, its accessories and the equipment used, ranging. Affords study of compass surveying and its principles, types, leveling principles and theory, adjustments and errors in leveling.

Unit II: Provides a detail study of theodolite, the horizontal and vertical measurements. Calculating height and distance. Provides an idea on Tachometric surveying. It gives an overview on contouring, its characteristics and methods.

Unit III: The basics of geodetic surveying, its measurement in both vertical and horizontal observations and its surveying adjustments are been discussed. Also it presents the details of the errors classifications.

Unit IV: This unit carries the information of advanced surveying topics that includes both hydrographic surveying and astronomical surveying the basics definition included in it, field observation and determination of time etc.. Are been discussed.

Unit V: Modern surveying techniques such as total station, GPS and its working principles. It affords the different segments involved in it.

Acknowledgement

The author Dr.G. Venkatesan thanks to the family members C. Govindaraj (Father), G. Krishnaveni (Mother), V. Sindhu Venkatesan (Wife), V. Haeshika (Daughter), V. Taniska (Daughter) for giving him, strength and determination to undertake and complete this surveying book successfully.

The author M. Aarthi thanks to the family members M. Murthy (Father), M. Jayalakshmi (Mother) and M. Chitra (Sister) for giving her, strength and determination to undertake and complete this surveying book successfully.

TABLE OF CONTENTS

UNIT I

1. FUNDAMENTALS OF CONVENTIONAL SURVEYING AND LEVELLING

Classifications and basic principles of surveying - Equipment and accessories for ranging and chaining - Methods of ranging - Compass - Types of Compass - Basic Principles - Bearing - Types - True Bearing - Magnetic Bearing - Levelling - Principles and theory of Levelling - Datum - Bench Marks - Temporary and Permanent Adjustments - Methods of Levelling - Booking Reduction - Sources of errors in Levelling - Curvature and refraction.

1.1. Surveying

The practice of measuring angles and distances on the ground so that they can be accurately plotted on a map.

1.2. Principles of Surveying

The fundamental principles upon which the surveying is being carried out are:

- Working from whole to part.
- After deciding the position of any point, its reference must be kept from at least two permanent objects or stations whose position have already been well defined.

The purpose of working from whole to part is,

- To localize the errors.
- To control the accumulation of errors.

1.3. Classifications of Surveying

Based on the purpose (for which surveying is being conducted), Surveying has been classified into:

Control surveying: To establish horizontal and vertical positions of control points.

Land surveying: To determine the boundaries and areas of parcels of land, also known as property survey, boundary survey or cadastral survey.

Topographic survey: To prepare a plan/ map of a region which includes natural as well as and man-made features including elevation.

Engineering survey: To collect requisite data for planning, design and execution of engineering projects. Three broad steps are:

1) Reconnaissance survey: To explore site conditions and availability of infrastructures.

2) Preliminary survey: To collect adequate data to prepare plan / map of area to be used for planning and design.

3) Location survey: To set out work on the ground for actual construction / execution of the project.

Route survey: To plan, design, and laying out of route such as highways, railways, canals, pipelines, and other linear projects.

Construction surveys: Surveys which are required for establishment of points, lines, grades, and for staking out engineering works (after the plans have been prepared and the structural design has been done).

Astronomic surveys: To determine the latitude, longitude (of the observation station) and azimuth (of a line through observation station) from astronomical observation.

Mine surveys: To carry out surveying specific for opencast and underground mining purposes.

2. Chain Survey

Chain survey is the simplest method of surveying. In this survey only measurements are taken in the field, and the rest work, such as plotting calculation etc. are done in the office. This is most suitable adapted to small plane areas with very few details. If carefully done, it gives quite accurate results. The necessary requirements for field work are chain, tape, ranging rod, arrows and some time cross staff.

1.3.1. Survey Station

Survey stations are of two kinds.

    1. Main Stations.

    2. Subsidiary or tie.

Main Stations

Main stations are the end of the lines, which command the boundaries of the survey, and the lines joining the main stations re called the main survey line or the chain lines.

Subsidiary or the Tie Stations

Subsidiary or the tie stations are the point selected on the main survey lines, where it is necessary to locate the interior detail such as fences, hedges, building etc.

Tie or Subsidiary Lines

A tie line joints two fixed points on the main survey lines. It helps to checking the accuracy of surveying and to locate the interior details. The position of each tie line should be close to some features, such as paths, building etc.

Base Lines

It is main and longest line, which passes approximately through the centre of the field. All the other measurements to show the details of the work are taken with respect of this line.

Check Line

A check line also termed as a proof line is a line joining the apex of a triangle to some fixed points on any two sides of a triangle. A check line is measured to check the accuracy of the framework. The length of a check line, as measured on the ground should agree with its length on the plan.

Offsets

These are the lateral measurements from the base line to fix the positions of the different objects of the work with respect to base line. These are generally set at right angle offsets. It can also be drawn with the help of a tape. There are two kinds of offsets.

1) Perpendicular offsets.

2) Oblique offsets.

The measurements are taken at right angle to the survey line called perpendicular or right angled offsets.

The measurements which are not made at right angles to the survey line are called oblique offsets or tie line offsets.

1.3.2. Procedure in Chain Survey

1. Reconnaissance

The preliminary inspection of the area to be surveyed is called reconnaissance. The surveyor inspects the area to be surveyed, survey or prepares index sketch or key plan.

2. Marking Station

Surveyor fixes up the required no stations at places from where maximum possible stations are possible.

3. Then he selects the way for passing the main line, which should be horizontal and clean as possible and should pass approximately through the centre of work.

4. Then ranging roads are fixed on the stations.

5. After fixing the stations, chaining could be started.

6. Make ranging wherever necessary.

7. Measure the change and offset.

1.4. Classification of Surveying

Generally, surveying is divided into two major categories: plane and geodetic surveying.

1.4.1. Plane Surveying

PLANE SURVEYING is a process of surveying in which the portion of the earth being surveyed is considered a plane. The term is used to designate survey work in which the distances or areas involved are small enough that the curvature of the earth can be disregarded without significant error. In general, the term of limited extent. For small areas, precise results may be obtained with plane surveying methods, but the accuracy and precision of such results will decrease as the area surveyed increases in size. To make computations in plane surveying, you will use formulas of plane trigonometry, algebra, and analytical geometry.

A great number of surveys are of the plane surveying type. Surveys for the location and construction of highways and roads, canals, landing fields, and railroads are classified under plane surveying. When it is realized that an arc of 10 mi is only 0.04 greater that its subtended chord; that a plane surface tangent to the spherical arc has departed only about 8 in. at 1 mi from the point of tangency; and that the sum of the angles of a spherical triangle is only 1 sec greater than the sum of the angles of a plane triangle for a triangle having an area of approximately 75 sq mi on the earth's surface, it is just reasonable that the errors caused by the earth's considered curvature only in precise surveys be of large areas.

In this training manual, we will discuss primarily the methods used in plane surveying rather than those used in geodetic surveying.

1.4.2. Geodetic Surveying

GEODETIC SURVEYING is a process of surveying in which the shape and size of the earth are considered. This type of survey is suited for large areas and long lines and is used to find the precise location of basic points needed for establishing control for other surveys. In geodetic surveys, the stations are normally long distances apart, and more precise instruments and surveying methods are required for this type of surveying than for plane surveying.

The shape of the earth is thought of as a spheroid, although in a technical sense, it is not really a spheroid. In 1924, the convention of the International Geodetic and Geophysical Union adopted 41,852,960 ft as the diameter of the earth at the equator and 41,711,940 ft as the diameter at its polar axis. The equatorial diameter was computed on the assumption that the flattening of the earth caused by gravitational attraction is exactly 1/297.

Therefore, distances measured on or near the surface of the earth are not along straight lines or planes, but on a curved surface.

Hence, in the computation of distances in geodetic surveys, allowances are made for the earth's minor and major diameters from which a spheroid of reference is developed. The position of each geodetic station is related to this spheroid. The positions are expressed as latitudes (angles north or south of the Equator) and longitudes (angles east or west of a prime meridian) or as northings and castings on a rectangular grid.

The methods used in geodetic surveying are beyond the scope of this training manual.

Topographic Surveys

The purpose of a TOPOGRAPHIC SURVEY is to gather survey data about the natural and man-made features of the land, as well as its elevations. From this information a three-dimensional map may be prepared. You may prepare the topographic map in the office after collecting the field data or prepare it right away in the field by plane table. The work usually consists of the following:

1. Establishing horizontal and vertical control that will serve as the framework of the survey.
2. Determining enough horizontal location and elevation (usually called side shots) of ground points to provide enough data for plotting when the map is prepared.
3. Locating natural and man-made features that may be required by the purpose of the survey.
4. Computing distances, angles, and elevations.
5. Drawing the topographic map.

Topographic surveys are commonly identified with horizontal and/or vertical control of third- and lower-order accuracies.

Route Surveys

The term *route survey* refers to surveys necessary for the location and construction of lines of transportation or communication that continue across country for some distance, such as highways, railroads, open-conduit systems, pipelines, and power lines. Generally, the preliminary survey for this work takes the form of a topographic survey. In the final stage, the work may consist of the following:

1. Locating the center line, usually marked by stakes at 100-ft intervals called stations.
2. Determining elevations along and across the center line for plotting profile and cross sections.
3. Plotting the profile and cross sections and fixing the grades.
4. Computing the volumes of earthwork and preparing a mass diagram.
5. Staking out the extremities for cuts and fills.
6. Determining drainage areas to be used in the design of ditches and culverts.
7. Laying out structures, such as bridges and culverts.
8. Locating right-of-way boundaries, as well as staking out fence lines, if necessary.

Special Surveys

As mentioned earlier in this chapter, SPECIAL SURVEYS are conducted for a specific purpose and with a special type of surveying equipment and methods. A brief discussion of some of the special surveys familiar to you follows.

Land Surveys

LAND SURVEYS (sometimes called cadastral or property surveys) are conducted to establish the exact location, boundaries, or subdivision of a tract of land in any specified area. This type of survey requires professional registration in all states. Presently, land surveys generally consist of the following chores:

1. Establishing markers or monuments to define and thereby preserve the boundaries of land belonging to a private concern, a corporation, or the government.
2. Relocating markers or monuments legally established by original surveys. This requires examining previous survey records and retracing what was done. When some markers or monuments are missing, they are re-established following recognized procedures, using whatever information is available.
3. Rerunning old land survey lines to determine their lengths and directions. As a result of the high cost of land, old lines are re-measured to get more precise measurements.

4. Subdividing landed estates into parcels of predetermined sizes and shapes.

5. Calculating areas, distances, and directions and preparing the land map to portray the survey data so that it can be used as a permanent record. 6. Writing a technical description for deeds.

Control Surveys

CONTROL SURVEYS provide "basic control" or horizontal and vertical positions of points to which supplementary surveys are adjusted. These types of surveys (sometimes termed and traverse stations and the elevations of bench marks. These control points are further used as References for hydrographic surveys of the coastal waters; for topographic control; and for the control of many state, city, and private surveys.

Horizontal and vertical controls generated by land (geodetic) surveys provide coordinated position data for all surveyors. It is therefore necessary that these types of surveys use first-order and second-order accuracies.

Hydrographic Surveys

HYDROGRAPHIC SURVEYS are made to acquire data required to chart and/or map shorelines and bottom depths of streams, rivers, lakes, reservoirs, and other larger bodies of water. This type of survey is also of general importance to navigation and to development of water resources for flood control, irrigation, electrical power, and water supply. As in other special surveys, several different types of electronic and radio-acoustical instruments are used in hydrographic surveys. These special devices are commonly used in determining water depths and location of objects on the bottom by a method called taking SOUNDINGS. Soundings are taken by measuring the time required for sound t to travel downward and be reflected back to a receiver aboard a vessel.

1.5. Types of Surveying Operations

The practice of surveying actually boils down to fieldwork and office work. The FIELDWORK consists of taking measurements, collecting engineering data, and testing materials. The OFFICE WORK includes taking care of the computation and drawing the necessary information for the purpose of the survey.

Fieldwork

FIELDWORK is of primary importance in all types of surveys. To be a skilled surveyor, you must spend a certain amount of time in the field to acquire needed experience. The study of this training manual will enable you to understand the underlying theory of surveying, the instruments and their uses, and the surveying methods. However, a high degree of proficiency in actual surveying, as in other professions, depends largely upon the duration, extent, and variation of your actual experience.

You should develop the habit of STUDYING the problem thoroughly before going into the field, You should know exactly what is to be done; how you will do it; why you prefer a certain approach over other possible solutions; and what instruments and materials you will need to accomplish the project.

It is essential that you develop SPEED and CONSISTENT ACCURACY in all your fieldwork. This means that you will need practice in handling the instruments, taking observations and keeping field notes, and planning systematic moves.

It is important that you also develop the habit of CORRECTNESS. You should not accept any measurement as correct without verification. Verification, as much as possible, should be different from the original method used in measurement. The precision of measurement must be consistent with the accepted standard for a particular purpose of the survey.

Fieldwork also includes adjusting the instruments and caring for field equipment. Do not attempt to adjust any instrument unless you understand the workings or functions of its parts. Adjustment of instruments in the early stages of your career requires close supervision from a senior EA.

Factors Affecting Fieldwork

The surveyor must constantly be alert to the different conditions encountered in the field. Physical factors, such as TERRAIN AND WEATHER CONDITIONS, affect each field survey in varying degrees. Measurements using telescopes can be stopped by fog or mist. Swamps and flood plains under high water can impede taping surveys. Sights over open water or fields of flat, unbroken terrain create ambiguities in measurements using microwave equipment. The lengths of light-wave distance in measurements are reduced in bright sunlight. Generally, reconnaissance will predetermine the conditions and alert the survey party to the best method to use and the rate of progress to expect.

The STATE OF PERSONNEL TECHNICAL READINESS is another factor affecting field-work. As you gain experience in handling various surveying instruments, you can shorten survey time and avoid errors that would require resurvey.

The PURPOSE AND TYPE OF SURVEY are primary factors in determining the accuracy requirements. First-order triangulation, which becomes the basis or "control" of future surveys, is made to high-accuracy standards. At the other extreme, cuts and fills for a highway survey carry accuracy standards of a much lower degree. In some construction surveys, normally inaccessible distances must be computed. The distance is computed by means of trigonometry, using the angles and the one distance that can be measured. The measurements must be made to a high degree of precision to maintain accuracy in the computed distance.

So, then, the purpose of the survey determines the accuracy requirements. The required accuracy, in turn, influences the selection of instruments and procedures. For instance, comparatively rough procedures can be used in measuring for earthmoving, but grade and alignment of a highway have to be much more precise, and they, therefore, require more accurate measurements. Each increase in precision also increases the time required to make the measurement, since greater care and more observations will be taken. Each survey measurement will be in error to the extent that no measurement is ever exact. The errors are classified as systematic and accidental and are explained in the latter part of this text. Besides errors, survey measurements are subject to mistakes or blunders. These arise from misunderstanding of the problem, poor judgment, confusion on the part of the surveyor, or simply from an oversight. By working out a systematic procedure, the surveyor will often detect a mistake when some operation seems out of place. The procedure will be an advantage in setting up the equipment, in making observations, in recording field notes, and in making computations.

Survey speed is not the result of hurrying; it is the result of saving time through the following factors:

1. The skill of the surveyor in handling the instruments.
2. The intelligent planning and preparation of the work.
3. The process of making only those measurements that are consistent with the accuracy requirements.

Experience is of great value, but in the final analysis, it is the exercise of a good, mature, and competent degree of common sense that makes the difference between a good surveyor and an exceptional surveyor.

Field Survey Parties

The size of a field survey party depends upon the survey requirements, the equipment available, the method of survey, and the number of personnel needed for performing the different functions. Four typical field survey parties commonly used in the SEABEEs are briefly described in this section: a level party, a transit party, a stadia party, and a plane table party.

LEVEL PARTY: The smallest leveling party consists of two persons: an instrument man and a rodman. This type of organization requires the instrument man to act as note keeper. The party may need another recorder and one or more extra rodmen to improve the efficiency of the different leveling operations. The addition of the rodmen eliminates the waiting periods while one person moves from point to point, and the addition of a recorder allows the instrument man to take readings as soon as the rodmen are in position. When leveling operations are run along with other control surveys, the leveling party may be organized as part of a combined party with personnel assuming dual duties, as required by the work load and as designated by the party chief.

TRANSIT PARTY: A transit party consists of at least three people: an instrument man, a head chainman, and a party chief. The party chief is usually the note keeper and may double as rear chainman, or there may be an additional rear chainman. The instrument man operates the transit; the head chainman measures the horizontal distances; and the party chief directs the survey and keeps the notes.

STADIA PARTY: A stadia party should consist of three people: an instrument man, a note keeper, and a rodman. However, two rodmen should be used if there are long distances between observed points so that one can proceed to a new point, while the other is holding the rod on a point being observed. The note keeper records the data called off by the instrument man and makes the sketches required.

PLANE TABLE PARTY: The plane table party consists of three people: a topographer or plane table operator, a rodman, and a computer. The topographer is the chief of the party who sets up, levels, and orients the plane table; makes the necessary readings for the determination of horizontal distances and elevations; plots the details on the plane table sheet as the work proceeds; and directs the other members of the party.

The rodman carries a stadia rod and holds it vertically at detail points and at critical terrain points in the plotting of the map. An inexperienced rodman must be directed by the topographer to each point at which the rod is to be held. An experienced rodman will expedite the work of the party by selecting the proper rod positions and by returning at times to the plane table to draw in special details that he may have noticed.

The computer reduces stadia readings to horizontal and vertical distances and computes the ground elevation for rod observations. He carries and positions the umbrella to shade the plane table and performs other duties as directed by the topographer. At times, the computer may be used as a second rodman, especially when the terrain is relatively flat and computations are mostly for leveling alone.

Field Notes

Field notes are the only record that is left after the field survey party departs the survey site. If these notes are not clear and complete, the field survey was of little value. It is therefore necessary that your field notes contain a complete record of all of the measurements made during the survey and that they include, where necessary, sketches and narrations to clarify the notes. The following guidelines apply.

LETTERING: All field notes should be lettered legibly. The lettering should be in freehand, vertical or slanted Gothic style, as illustrated in basic drafting. A fairly hard pencil or a mechanical lead holder with a 3H or 4H lead is recommended. Numerals and decimal points should be legible and should permit only one interpretation.

FORMAT: Notes must be kept in the regular field note book and not on scraps of paper for later transcription. Separate surveys should be recorded on separate pages or in different books. The front cover of the field notebook should be marked with the name of the project, its general location, the types of measurements recorded, the designation of the survey unit, and other pertinent information.

The inside front cover should contain instructions for the return of the notebook, if lost. The right-hand pages should be reserved as an index of the field notes, a list of party personnel and their duties, a list of the instruments used, dates and reasons for any instrument changes during the course of the survey, and a sketch and description of the project.

Throughout the remainder of the notebook, the beginning and ending of each day's work should be clearly indicated. Where pertinent, the weather, including temperature and wind velocities, should also be recorded. To minimize recording errors, someone other than the recorder should check and initial all data entered in the notebook.

RECORDING: Field note recording takes three general forms: tabulation, sketches, and descriptions. Two, or even all three, forms may be combined, when necessary, to make a complete record.

In TABULATION, the numerical measurements are recorded in columns according to a prescribed plan. Spaces are also reserved to permit necessary computations.

SKETCHES add much to clarify field notes and should be used liberally when applicable. They may be drawn to an approximate scale, or important details may be exaggerated for clarity. A small ruler or triangle is an aid in making sketches. Measurements should be added directly on the sketch or keyed in some way to the tabular data. An important requirement of a sketch is legibility. See that the sketch is drawn clearly and large enough to be understandable.

Tabulation, with or without added sketches, can also be supplemented with DESCRIPTIONS. The description may be only one or two words to clarify t he recorded measurements. It may also be quite a narration if it is to be used at some future time, possibly years later, to locate a survey monument.

MESURES ARE NOT PERMITTED IN FIELD NOTEBOOKS. Individual numbers or lines recorded incorrectly are to be lined out and the correct values inserted. Pages that are to be rejected are crossed out neatly and referenced to the substituted pages. THIS PROCEDURE IS MANDATORY since the field notebook is the book of record and is often used as legal evidence. Standard abbreviations, signs, and symbols are used in field notebooks. If there is any doubt as to their meaning, an explanation must be given in the form of notes or legends.

Office Work

OFFICE WORK in surveying consists of converting the field measurements into a usable format. The conversion of computed, often mathematical, values may be required immediately to continue the work, or it may be delayed until a series of field measurements is completed. Although these operations are performed in the field during lapses between measurements, they can also be considered office work. Such operations are normally done to save time. Special equipment, such as calculators, conversion tables, and some drafting equipment, are used in most office work.

In office work, converting field measurements (also called reducing) involves the process of computing, adjusting, and applying a standard rule to numerical values.

Computation

In any field survey operation, measurements are derived by the application of some form of mathematical computation. It may be simple addition of several full lengths and a partial tape length to record a total linear distance between two points. It maybe the addition or subtraction of differences in elevation to determine the height of instrument or the elevation during leveling. Then again, it may be checking of angles to ensure that the allowable error is not exceeded.

Office computing converts these distances, elevations, and angles into a more usable form. The finished measurements may end up as a computed volume of dirt to be moved for a highway cut or fill, an area of land needed for a SEABEE construction project, or a new position of a point from which other measurements can be made.

In general, office computing reduces the field notes to either a tabular or graphic form for a permanent record or for continuation of fieldwork.

Adjustment

Some survey processes are not complete until measurements are within usable limits or until corrections have been applied to these measurements to distribute accumulated errors. Small errors that are not apparent in individual measurements can accumulate to a sizeable amount. Adjusting is the process used to distribute these errors among the many points or stations until the effect on each point has been reduced to the degree that all measurements are within usable limits.

For example, assume that 100 measurements were made to the nearest unit for the accuracy required. This requires estimating the nearest one-half unit during measurement. At the end of the course, an error of + 4 units results. Adjusting this means each measurement is reduced 0.04 unit. Since the measurements were read only to the nearest unit, this adjustment would not be measurable at any point, and the adjusted result would be correct.

Significant Figures

The term known to be exact.

In a measured quantity, the number of significant figures is determined by the accuracy of the measurement. For example, a roughly measured distance of 193 ft has three significant figures. More carefully measured, the same distance, 192.7 ft, has four significant figures. If measured still more accurately, 192.68 ft has five significant figures.

In surveying, the significant figures should reflect the allowable error or tolerance in the measurements. For example, suppose a measurement of 941.26 units is made with a probable error of 0.03 unit. The 0.03 casts some doubt on the fifth digit which can vary from 3 to 9, but the fourth digit will still remain 2. We can say that 941.26 has five significant figures; and from the allowable error, we know the fifth digit is doubtful.

However, if the probable error were 0.07, the fourth digit could be affected. The number could vary from 941.19 to 941.33, and the fourth digit could be read 1, 2, or 3. The fifth digit in this measurement is meaningless. The number has only four significant figures and should be written as such.

The number of significant figures in a number ending in one or more zeros is unknown unless more information is given. The zeros may have been added to show the location of the decimal point; for example, 73200 may have three, four, or five significant figures, depending on whether the true value is accurate to 100, 10, or 1 unit(s). If the number is written 73200.0, it indicates accuracy is carried to the tenth of a unit and is considered to have six significant figures.

When decimals are used, the number of significant figures is not always the number of digits. A zero may or may not be significant, depending on its position with respect to the decimal and the digits. As mentioned above, zeros may have been added to show the position of the decimal point. Study the following examples:

0.000047........two significant figures

0.0100470........six significant figures

0.1000470........seven significant figures

2.0100470........eight significant figures

In long computations, the values are carried out to one more digit than required in the result. The number is rounded off to the required numbers of digits as a final step.

ROUNDING OFF NUMBERS: Rounding off is the process of dropping one or more digits and replacing them with zeros, if necessary, to indicate the number of significant figures. Numbers used in surveying are rounded off according to the following rules:

1. When the digit to be dropped is less than 5, the number is written without the digit or any others that follow it. (Example: 0.054 becomes 0.05.)

2. When the digit is equal to 5, the nearest EVEN number is substituted for the preceding digit. (Examples: 0.055 becomes 0.06; 0.045 becomes 0.04.)

3. When the digit to be dropped is greater than 5, the preceding digit is increased by one. (Example: 0.047 becomes 0.05.)

4. Dropped digits to the left of the decimal point are replaced by zeros.

5. Dropped digits to the right of the decimal points are never replaced.

Examples

2738.649 to five significant figures equals 2738.6

792.850 to five significant figures equals 792.8

792.750 to five significant figures equals 792.8

675823. to five significant figures equals 675800

675863 to five significant figures equals 675900

4896.3 to five significant figures equals 4896

4896.7 to five significant figures equals 4897

Checking Computations

Most mathematical problems can be solved by more than one method. To check a set of computations, you should use a method that differs from the original method, if possible. An inverse solution, starting with the computed value and solving for the field data, is one possibility. The planimeter and the protractor are also used for approximate checking. A graphical solution can be used, when feasible, especially if it takes less time than a mathematical or logarithmic solution. Each step that cannot be checked by any other method must be recomputed; and, if possible, another EA should re-compute the problem. When an error or mistake is found, the computation should be rechecked before the correction is accepted.

Drafting Used in Surveying

The general concept of drafting and the use of drafting instruments were discussed in chapters 2 through 5. By this time, you should be familiar with the use of various drafting instruments and with the elements of mechanical drawing. Drafting used in surveying, except for some freehand sketches, is generally performed by mechanical means; for example, the drawing of lines and surveying symbols is generally done with the aid of a straightedge, spline, template, and so on.

The drawings you make that are directly related to surveying will consist of maps, profiles, cross sections, mass diagrams, and, to some extent, other graphical calculations. Their usefulness depends upon how accurately you plot the points and lines representing the field measurements. It is important that you adhere to the requirements of standard drawing practices. Correctness, neatness, legibility, and well proportioned drawing arrangements are signs of professionalism.

In drawing a PROPERTY map, for example, the following general information must be included:

1. The length of each line, either indicated on the line itself or in a tabulated form, with the distances keyed to the line designation.

2. The bearing of each line or the angles between lines.

3. The location of the mapped area as referenced to an established coordinate system.

4. The location and kind of each established monument indicating distances from reference marks.

5. The name of each road, stream, landmark, and so on.

6. The names of all property owners, including those whose lots are adjacent to the mapped area.

7. The direction of the true or magnetic meridian, or both.

8. A graphical scale showing the corresponding numerical equivalent.

9. A legend to the symbols shown on the map, if those shown are not standard signs.

10. A title block that distinctly identifies the tract mapped or the owner's map. (It is required to contain the name of the surveyor, the name of the draftsman, and the date of the survey.)

Besides the above information, there are some other items that may be required if the map is to become a public record. When this is the case, consult the local office of the Bureau of Land Management or the local surveyors' society for the correct general information requirements to be included in the map to be drawn.

In drawing maps that will be used as a basis for studies, such as those to be used in roads, structures, or waterfront construction, you are required to include the following general information:

1. Information that will graphically represent the features in the plan, such as streams, lakes, boundaries, roads, fences, and condition and culture of the land.

2. The relief or contour of the land.

3. The graphical scale.

4. The direction of the meridian.

5. The legend to symbols used, if they are not conventional signs.

6. A standard title block with a neat and appropriate title that states the kind or purpose of the map. Again, the surveyor's name and that of the draftsman, as well as the date of survey, are to be included in the title block.

Maps developed as a basis for studies are so varied in purpose that the above information may be adequate for some but inadequate for others. The Engineering Aid, when in doubt, should consult the senior EA, the engineering officer, or the operations officer as to the information desired in the proposed map. The senior EA or the chief of the field survey party is required to know all these requirements before actual fieldwork is started.

A map with too much information is as bad as a map with too little information on it. It is not surprising to find a map that is so crowded with information and other details that it is hard to comprehend. If this happens, draw the map to a larger scale or reduce the information or details on it. Then, provide separate notes or descriptions for other information that will not fit well and thus will cause the appearance of overcrowding. Studying the features and quality of existing maps developed by NAVFACENGCOM and civilian architects and engineers (A & E) agencies will aid you a great deal in your own map drawing.

1.6. Ranging and Chaining

The process of fixing or establishing intermediate points is known as ranging. Two methods of ranging:

- Direct ranging.
- Indirect ranging.

1.6.1. Direct Ranging

1. Direct ranging is done when the two ends of the survey lines are intervisible. In such case, ranging can either be done by eye or through some optical instrument such as a line ranger or a theodolite.
2. Let A & B be the two points at the ends of a survey line. One ranging rod is erected at the point B while the surveyor stands with another ranging rod at point A holding the rod at about half metre length.
3. The assistant then goes with another ranging rod and establishes the rod at a point approximately in the line with AB at a distance not greater than one chain length from A.
4. The surveyor at A then signals the assistant to move transverse to the chain line, till he is in line with A & B.

Ranging by Line Ranger

1. A line ranger consists of either two plane mirrors or two right angled isosceles prisms one above the other.
2. The diagonals of the two prisms are silvered so as to reflect the incidental rays.
3. A handle with a hook is provided at the bottom to hold the instrument in hand to transfer the point on the ground wit the help of plumb-bob.

4. To range a point P, two ranging rods are fixed at the ends A &B, the surveyor at P holds the line ranger very near to the line AB.

5. The lower prism abc receives the rays from A which are reflected by the diagonal ac towards the observer.

6. Similarly, the upper prism dbc receives the rays from B which are reflected by the diagonal bd towards the observer. Thus, the observer views the images of ranging rods at A & B.

7. The surveyor then moves the instrument sideways till the two images are in the same vertical line.

8. The point P is then transferred to the ground with the help of a plumb bob.

1.6.2. Indirect or Reciprocal Ranging

Indirect or reciprocal ranging is resorted to when both the ends of the survey line are not intervisible either due to high intervening ground or due to long distance between them. In such case, ranging is done indirectly by selecting two intermediate points M1 and N1 very near to the chain line in such a way that from M1 both N1 are visible and from N1, both M1 and A are visible.

1. Two surveyors station themselves at M1 and N1 with ranging rods. The person at M1 then directs the person at N1 to move to a new position N2 in line with M1B.

2. The person at N2 then directs the person at M1 to move to a new position M2 in line N2A. Thus, the two persons are now at M2 and N2 which are nearer to the chain line than the positions M1 and N1.

3. The process is repeated till the points M and N are located in such a way that the person at M finds the person at N in line with MB, and the person at N finds the person at M in line with NA.

4. After having established M & N, other points can be fixed by direct ranging.

1.6.3. Chaining

Two chainmen are required for measuring the length of a line which is greater than a chain length.

Follower: The more experienced of the chainmen remains at the zero end or rear end of the chain and is called the follower.

Leader: The other chainmen holding the forward handle is known as the leader.

Unfolding the Chain

1. To unfold the chain, the chainmen keeps both the handles in the left hand and throws the rest of the portion of the chain in the forward direction with his right hand.
2. The other chainmen assists in removing the knots etc. and in making the chain straight.

1.6.4. Lining and Marking

1. The follower holds the zero end of the chain at the terminal point while the leader proceeds forward with the other end in one hand and a set of 10 arrows and a ranging rod in the other hand.
2. When he is approximately one chain length away, the follower directs him to fix his pole in line with the pole.
3. When the point is ranged, the leader makes a mark on the ground, holds the handle with both the hands and pulls the chain so that it becomes straight between the terminal point and the point fixed.
4. Little jerks given for his purpose but the pull applied must be just sufficient to make the chain straight in line.
5. The leader then puts an arrow at the end of the chain, swings the chain slightly out of the line and proceeds further with the handle in one hand and the rest of the arrows and ranging rod in the other hand.
6. The follower also takes the end handle in one hand and a ranging rod in the other hand, follows the leader till the leader has approximately traveled one chain length.
7. The follower puts the zero end of the chain at first arrow fixed by the leader, and ranges the leader who in turn, stretches the chain straight in the line & fixes the second arrow in the ground and proceeds the further.
8. The follower takes the first arrow and the ranging rod in one hand and the handle in the other & follows the leader.
9. At the end of ten chains, the leader calls for the 'arrows'.
10. The follower takes out the tenth arrow from the ground, puts a ranging rod there & hands over ten arrows to the leader.
11. The transfer of ten arrows is recorded by the surveyor.
12. To measure the fractional length at the end of a line, the leader drags the chain beyond the end station, stretches it straight and tight the reads the links.

1.6.5. Chaining on Uneven or Sloping Ground

Two methods:

1. Direct Method
2. Indirect Method

Direct Method

1. In the direct method the distance is measured in small horizontal stretches or steps.
2. For example to measure the distance between the 2 points A & B.
3. The follower holds the zero end of the tape at A while the leader selects any suitable length l_1 of the tape and moves forward. The follower directs the leader for ranging.
4. The leader pulls the tape tight, makes it horizontal and the point 1 is then transferred to the ground by a plumb bob.
5. A special form of drop arrow is used to transfer the point to the surface. The procedure is then repeated.
6. The total length D of the line is then equal to $(l_1+l_2+l_3\ldots\ldots)$. This method followed in case of irregular slopes.

Indirect Method

1. Angle Method

1. Let l_1 = measured inclined distance between AB and θ = slope of AB with horizontal. The horizontal distance D_1 is given by $D_1 = l_1\cos\theta_1$.
2. Similarly for BC, $D_2 = l_2\cos\theta_2$
3. The required horizontal distance between any two points $= \sum l \cos\theta$
4. The slopes of the lines can be measured with the help of a clinometer.
5. A clinometer, in its simplest form essentially consists of a line of sight, a graduated arc, a light plumb bob with a long thread suspended at the centre.

A plumb is suspended from C, the central point. When the clinometer is horizontal, the thread touches the zero mark of the calibrated circle. To sight a point, the clinometer is tilted so that the line of sight AB may pass through the object. Since the thread still remains vertical, the reading against the thread gives the slope of the line of sight.

2. Difference in Level Measured

1. Sometimes, in the place of measuring the angle θ, the difference in the level between the points is measured with the help of a leveling instrument and the horizontal

distance is computed.

2. Thus, if h is the difference in level, we have

$$D=\sqrt{l^2-h^2}$$

3. *Hypotenusal Allowance*

1) In this method, a correction is applied in the field at every chain length and at every point where the slope changes.

2) When the chain is stretched on the slope, the arrow is not put at the end of the chain but is placed in advance of the end, by an amount which allows for the slope correction.

3) BA' is one chain length slope. The arrow is not put A', the distance AA' being of such magnitude that the horizontal equivalent of BA is equal to 1 chain.

1.7. Traversing

Traversing is that type of survey in which a number of connected survey lines form the framework and the directions and lengths of the survey lines are measured with the help of an angle measuring instrument and a tape respectively.

Method of traversing:

- Chain traversing.
- Chain and compass traversing.
- Transit tape traversing:

    (a) By fast needle method.

    (b) By measurement of angles between the lines.

- Plane-table traversing.

1.8. Prismatic Compass

- Prismatic compass is the most convenient and portable of magnetic compass which can either be used as a hand instrument or can be fitted on a tripod.
- The magnetic needle is attached to the circular ring or compass card made up of aluminium, a non- magnetic substance.
- When the needle is on the pivot it will orient itself in the magnetic meridian and, therefore, the N and S ends of thc ring will be in this direction.
- The line of sight is defined by the objective vane and the eye slit, both attached to the compass box.

- The object vane consists of a vertical hair attached to a suitable frame while the eye slit consists of a vertical slit cut into the upper assembly of the prism unit, both being hinged to the box.

- When an object is sighted, the sight vanes wilt rotate with respect to the NS end of ring through an angle which the line makes with the magnetic meridian.

- A triangular prism is fitted below the eye slit having suitable arrangement for focusing to suit different eye sights. The prism has both horizontal and vertical faces convex, so that a magnified image of the ring graduation is formed. When the line of sight is also in the magnetic meridian, the South end ring comes vertically below the horizontal face of the prism.

- The 0°or 360° reading is, therefore, engraved on the South end of the ring, so that bearing of the magnetic meridian is read as 0°.

- The object vane presses against a bent lever which lifts the needle off the pivot and holds it against the glass lid.

- By pressing knob or break pin placed at the base of the object vane, a light spring fitted inside the box can be brought into the contact with the edge of the graduated ring to damp the oscillations of the needle when about to take the reading.

- The greatest advantage of prismatic compass is that both sighting he object as well as reading circle can be done simultaneously without hanging the position of the eye. The circle is read at the reading at which the hair line appears to cut the graduated ring.

1.8.1. Adjustment of Prismatic Compass

Station or Temporary Adjustments

Centring: Centring is the process of keeping the instrument exactly over the station.
Levelling: If the instrument is a hand instrument, it must in hand in such a way that graduated disc is swinging freely appears to be level as judged from the top edge of the ease.

Focusing the Prism: The prism attachment is slided up or down for focusing till the readings are seen to be sharp and clear.

Permanent Adjustments

- The permanent adjustments of prismatic compass are almost the same as that of the surveyor's except that there are no bubble tubes to be adjusted and the needle, cannot be straightened.

- The sight vanes are generally not adjustable.

The Surveyor's Compass

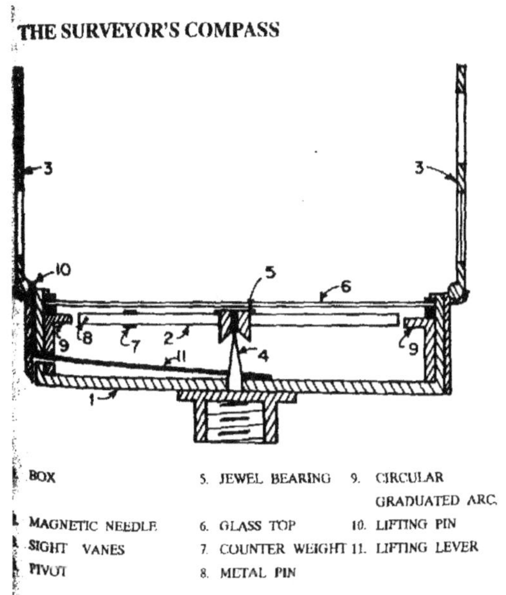

Figure 1.8.1: Shows the Essential Parts of a Surveyor's Compass

- The graduated ring is directly attached to the box and not with needle. The edge bar needle freely floats over the pivot.
- Thus, the graduated card or ring is not oriented in the magnetic meridian, as was the case in the prismatic compass.
- The object vane is similar to that of prismatic compass.
- The eye vane consists of a simple metal vane with a fine slit.
- Since no prism is provided, the object is to be sighted first with the object and eye vanes and the reading is then taken against the North end of the needle, by looking vertically through the top glass.
- When line of sight is in magnetic meridian, the North and south ends of the needle will be over the 0° N and 0° S graduations.

The card is graduated in quadrantal system having 0∘ at N and S ends and 90∘ East and West ends.

- Let us take the case of a line AB which is in North-East quadrant.
- In order to sight the point B, the box will have to be rotated about the vertical axis.

- In doing so, the pointer of the needle remains fixed in position while 0° N graduation of the card moves in a clockwise direction.

- Taking when the line has a bearing of 90° in East direction, the pointer appears to move by 90° from the 0° N graduation in anti-clockwise direction.

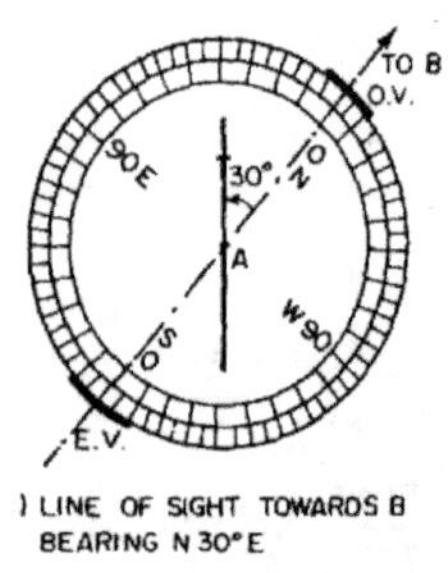

Figure 1.8.2

Difference between Prismatic Compass & Surveyor's Compass

Prismatic Compass	Surveyor's Compass
The needle is of 'broad needle' type. The needle does not act as index	The needle is of 'edge bar' type. The needle acts as index also.
i) The graduated card ring is attached with the needle. The ring does not rotate along with the line of sight.	i) The graduated card is attached to the box and not to the needle. The card rotates along with the line of sight.
ii) The graduations are in W.C.B system having 0º at south end, 90º at West, 180º at North and 270º at East.	ii) The graduations are in Q. B system, having 0º at N and S and 90º at East and West. East and West are interchanged.
iii) The graduations are engraved inverted.	iii) The graduations are engraved erect.
i) The object vane consists of metal vane with a vertical hair.	i) The object vane consists of a metal vane with a vertical hair.
ii) The eye vane consists of a small metal vane with slit.	ii) The eye vane consists of a metal vane with a fine slit.
i) The reading is taken with the help of a prism provided at the eye slit.	i) The reading is taken by directly seeing through the top of the glass.
ii) Sighting and reading taking can be done simultaneously from one position of the observer.	ii) Sighting and reading taking cannot be done simultaneously from one position of the observer.
Tripod may or may not be provided. The instrument can be used even by holding suitably in hand.	The instrument cannot be used without a tripod.

1.9. Bearing

Bearing of a line is its direction relative to a given meridian.

A meridian is any direction such as:

- True meridian.
- Magnetic Meridian.
- Arbitrary Meridian.

(1) True Meridian. The meridian through a point is the line in which a plane, passing that point and the north and south pd intersects with surface of the earth. It, thus, passes through the north and south.

True Bearing. True bearing of a line is the horizontal angle which it makes with the true meridian through one of the extremities of the line. Since the direction of true meridian through a point remains fixed, the true bearing of a line is a constant quantity.

(2) Magnetic Meridian Magnetic meridian through a is the direction shown by a freely floating and balanced magnetic needle free from all other attractive forces.

Magnetic Bearing The magnetic bearing of a line is the horizontal angle which it makes with the magnetic meridian passing through one of the extremities of the line.

(3) Arbitrary Meridian. Arbitrary meridian is any convenient direction towards a permanent and prominent mark or signal, such as a church spire or top of a chimney.

Arbitrary bearing Arbitrary bearing of a line is the horizontal angle which it makes with any arbitrary meridian passing through one of the extremities.

Conversion of W.C.B. into R.B

LINE	W.C.B	Rule for R.B.	Quadrant
AB	$0°$ & $90°$	R.B.=W.C.B	NE
AC	$90°$ & $180°$	R.B.= $180°$ -W.C.B	SE
AD	$180°$ & $270°$	R.B.=W.C.B.- $180°$	SW
AF	$270°$ & $360°$	R.B.= $360°$ -W.C.B.	NW

Conversion of R.B. into W.C.B

LINE	R.B.	Rule for W.C.B.	W.C.B between
AB	N α E	W.C.B= R.B	$0°$ & $90°$
AC	S β E	W.C.B.= $180°$ - R.B	$90°$ & $180°$
AD	S θ W	W.C.B.=$180°$ +R.B.	$180°$ & $270°$
AF	N ø W	W.C.B.= $360°$ - R.B	$270°$ & $360°$

1.9.1. Calculation of Angles from Bearings

- Knowing the bearing of two lines, the angle between the two can be easily calculated with the help of a diagrams.
- The included angle α between the lines AC and AB = θ_1 & θ_2= F.B. of one line— F.B. of the other line, both bearings being measured from a common point A. Refer Fig. the angle α = $(180° + \theta_1)$- θ_2= B. B. of Previous line. – F.B. of next line.

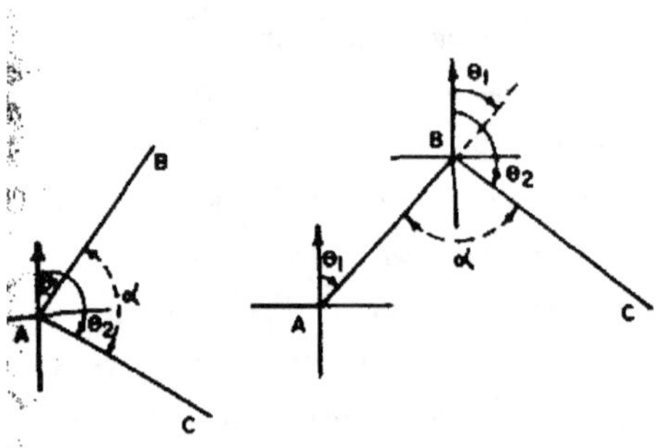

1.9.1

Let us consider the quadrantal bearing.

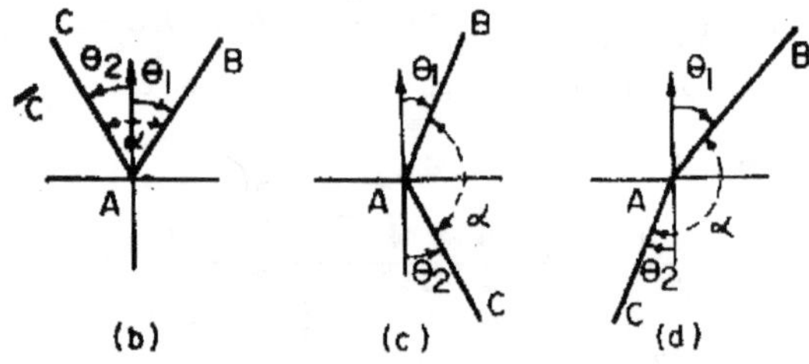

(b)   (c)   (d)

Figure 1.9.2: Quadrantal Bearing

- If the bearings have been measured to the same side of common meridian, the include angle $\alpha = \theta_2 - \theta_1$.
- In fig (b) both the bearings have been measured to the opposite sides of the common meridian, and included angle $\alpha = \theta_1 + \theta_2$.
- In fig © both the bearings have been measured to the same side of different and the included angle meridian, and included angle $\alpha = 180-(\theta_2 + \theta_1)$.

- In fig (d) both the bearings have been measured to the opposite sides of different meridians, and angles $\alpha = 180-(\theta_1- \theta_2)$.

1.9.2. Calculation of Bearings from Angles

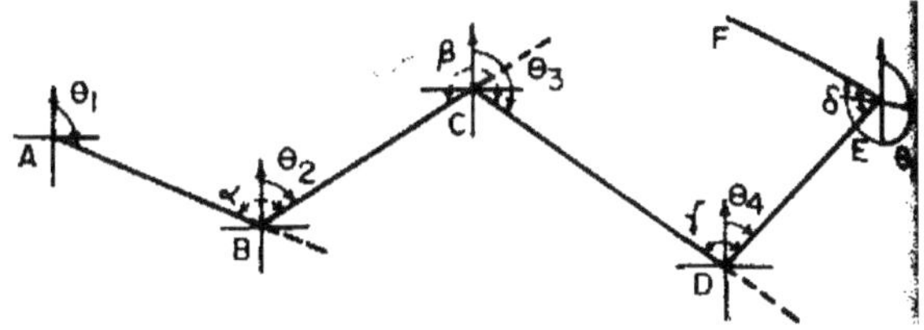

Figure 1.9.3

- Let α, β, γ, δ be the included angles measured clockwise from back stations & θ_1 be the measured bearing of the line AB.
- The bearing of the next line BC = θ_2 = θ_1 +α - 180
- The bearing of the next line CD = θ_3 = θ_2 + β - 180
- The bearing of the next line DE= θ_4= θ_3+ γ - 180
- The bearing of the next line EF= θ_5= θ_4+ δ +- 180
- (θ_1 +α), (θ_2 + β), (θ_3+ γ) are more than 180 while (θ_4+ δ) is less than 180.
- Therefore in order to calculate the bearing the following statement can be made:

"Add the measured clockwise angles to the bearing of the previous line. If the sum is more than 180, deduct 180°. If the sum is less than 180°, add 180°".

Examples on Angles and Bearings

Example: (a) Convert the following whole circle to quadrantal bearings,

    (i) 22°30'

    (ii) 170°12'

    (iii) 211°54'

    (iv) 327°24'.

(b) Convert the following quadrantal bearing to whole circle bearings,

    (i) N12°24'E

    (ii) S31°36'E

    (iii) S 68 6'W

    (iv) N5°42'W

Referring to fig above and tables given:

 (i) R.B.= W.CB =22°30'=N22°30'E

 (ii) R.B.= 180°—W. C. B. =180° - 170 12'=S 9° 48' E

 (iii) R.B.= W. C. B. — 180°=211° 54—180 °=S 31° 54'W

 (iv) R.B.= 360°—W.C.B.=360°—327° 24'=N 32° 36' W

 (i)WCB= RB=12°24'

 (ii) WCB = 180° — RB = 180° — 31° 36' =148° 24'

 (iii) W.C.B.= 180° + R.B.= 180° + 68° 6' = 248° 6'

 (iv) W.C.B.= 360° — R.B. = 360° — 5°42' = 354°18'

1.9.3. Earth's Magnetic Field and Dip

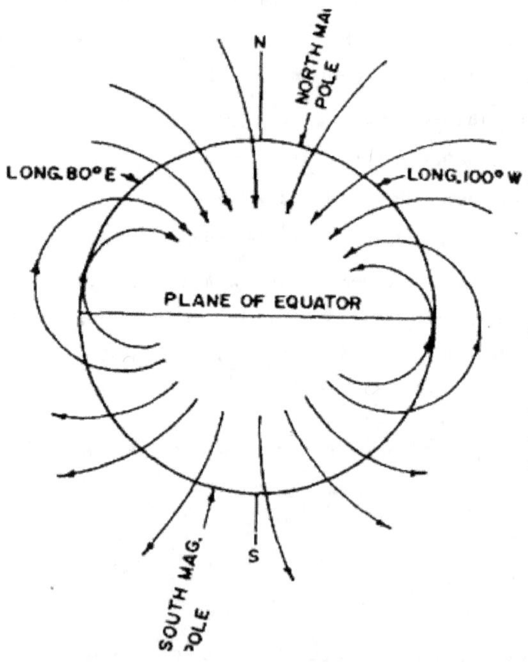

Figure 1.9.4 (a)

- The horizontal projections of the lines of force define the magnetic meridian. The angle which these lines of force make with the surface of the earth is called the **angle of dip or simply the dip of the needle.**

1.9.4. Magnetic Declination

- Magnetic declination at a place is the horizontal angle bet the true meridian and the magnetic meridian shown by the ne at the time of observation.
- If the magnetic meridian is to the right side (or eastern side) of the true meridian, declination is said to be eastern or positive, if it to be the left side (or western side), the declination is said to be western or negative.

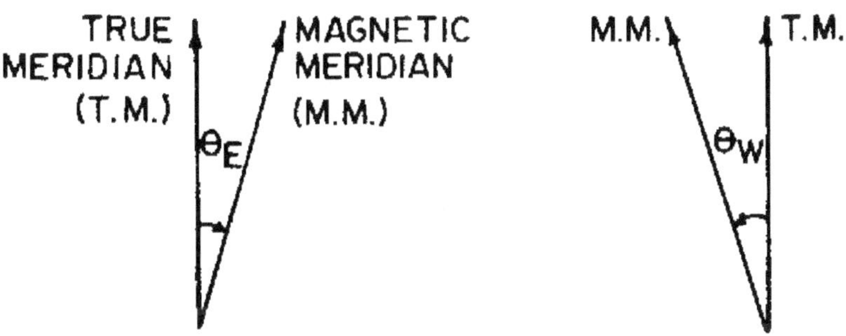

Figure 1.9.5: Magnetic Declination

1.10. Basic Surveying Instruments

Most fieldwork done by an Engineering Aid (especially at the third- and second-class levels) is likely to consist of field measurements and/or computations that involve plane surveying of ordinary precision. This section describes the basic instruments, tools, and other equipment used for this type of surveying. Other instruments used for more precise surveys will also be described briefly. Surveying instruments come in various forms, yet their basic functions are similar; that is, they are all used for measuring unknown angles and distances and/or for laying off known angles and distances.

1.10.1. Magnetic Compass

A magnetic compass is a device consisting principally of a circular compass card, usually graduated in degrees, and a magnetic needle, mounted and free to rotate on a pivot located at the center of the card. The needle, when free from any local attraction (caused by metal), lines itself up with the local magnetic meridian as a result of the attraction of the earth's magnetic North Pole.

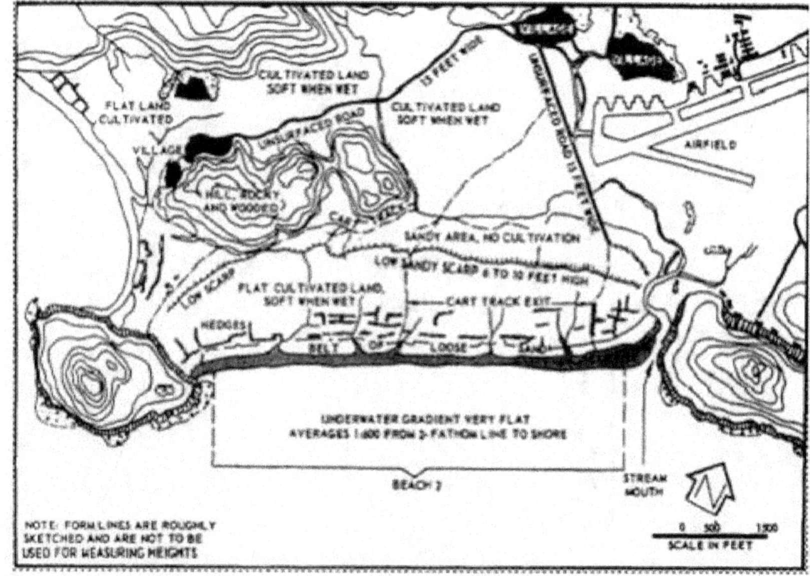

Figure 1.10.1: Line Map Made by Overlays from the Aerial Photograph

The magnetic compass is the most commonly used and simplest instrument for measuring directions and angles in the field. This instrument has a variety of both civilian and military applications. The LENSATIC COMPASS (available in your Table of Allowance) is most commonly used for SEABEE compass courses, for map orientation, and for angle direction during mortar and field artillery fires.

In addition to this type of compass, there are several others used exclusively for field surveys. The ENGINEER'S TRANSIT COMPASS, located between the standards on the upper plate, is graduated from 0° through 360° for measuring azimuths, and in quadrants of 90 ° for measuring bearings. The east and west markings are reversed. This permits direct reading of the magnetic direction.

The compass shown in figure 1.10.1 is commonly called the BRUNTON POCKET TRANSIT. This instrument is a combination compass and clinometer. It can be mounted on a light tripod or staff, or it may be cradled in the palm of the hand. Other types of compasses can also be found in some surveying instruments, such as the theodolite and plane table.

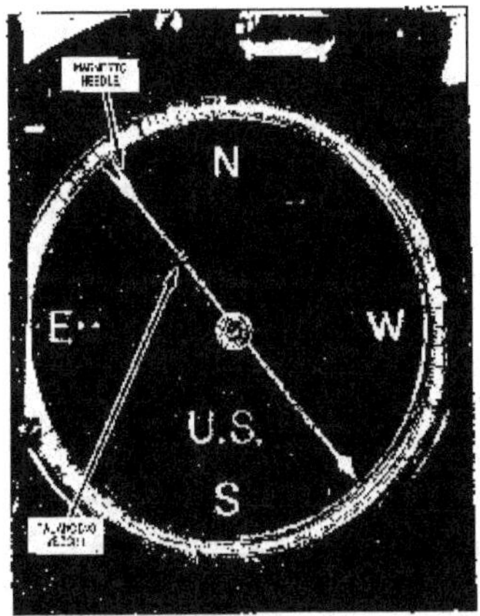

1.11. Engineer's Transit

A primary survey fieldwork consists of measuring horizontal and vertical angles or directions and extending straight lines. The instruments that can perform these functions have additional refinements (built-in) that can be used for other survey operations, such as leveling. Two types of instruments that fall into this category are the engineer' strans it and the theodolite. In recent years, manufacturing improvements have permitted construction of direct-reading theodolites that are soon to replace the vernier-reading transits. However, in most SEABEE construction, the engineer's sit tranisstill the major surveying instrument.

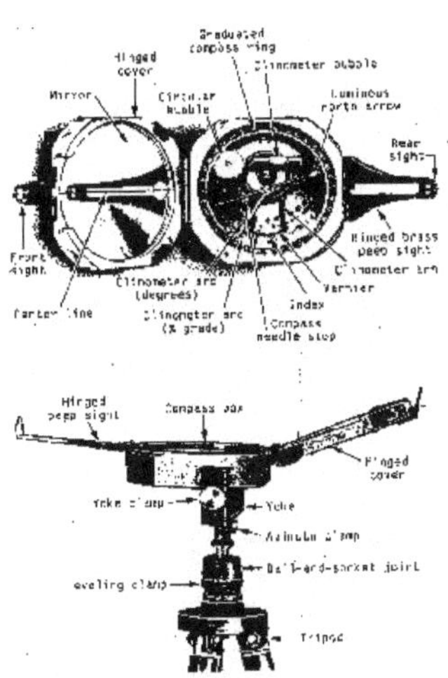

Figure 1.11.1: A Brunton Pocket Transit

The transit (fig. 1.11.1) is often called the universal survey instrument because of its uses. It may be used for measuring horizontal angles and directions, vertical angles, and differences in elevations; for prolonging straight lines; and for measuring distances by stadia. Although transits of various manufacturers differ in appearance, they are alike in their essential parts and operations. The engineer's transit contains several hundred parts. For-descriptive purposes, these parts may be grouped into three assemblies: the leveling head assembly, the lower plate assembly, and the upper many plate or alidade assembly.

Leveling Head Assembly

The leveling head of the transit normally is the four-screw type, constructed so the instrument can be shifted on the foot plate for centering over a marked point on the ground.

Lower Plate Assembly

The lower plate assembly of the transit consists of a hollow spindle that is perpendicular to center of a circular plate and accurately fitted the socket in the leveling head.

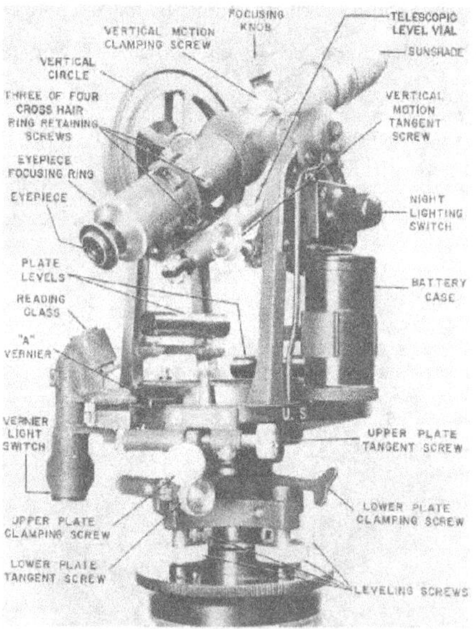

Figure 1.11.2: An Engineer's Transit

The lower plate contains the graduated horizontal circle on which the values of horizontal angles are read with the aid of two verniers, A and B, set on the opposite sides of the circle. A clamp controls the rotation of the lower plate and provides a means for locking it in place. A slow-motion tangent screw is used to rotate the lower plate a small amount to relative to the leveling head. The rotation accomplished by the use of the lower clamp and tangent screw is known as the LOWER MOTION.

Upper Plate or Alidade Assembly

The upper plate, alidade, or vernier assembly consists of a spindle attached plate to a circular plate carrying verniers, telescope standards, plate-level vials, and a magnetic compass. The spindle is accurately fitted to coincide with the socket in the lower plate spindle.

A clamp is tightened to hold the two plates together or loosened to permit the upper plate to rotate relative to the lower plate. A tangent screw permits the upper plate to be moved a small amount and is known as the UPPER MOTION.

The standards support two pivots with adjustable bearings that hold the horizontal axis and permit the telescope to move on a vertical plane. The vertical circle moves with the telescope. A clamp and tangent screw are provided to control this vertical movement. The vernier for the vertical circle is attached to the left standard. The telescope is an erecting type and magnifies the image about 18 to 25 times. The reticle contains stadia hairs in addition to the cross hairs. A magnetic compass is mounted on the upper plate between the two standards and consists of a magnetized needle pivoted on a jeweled bearing at the center of a graduated circle. A means is provided for lifting the needle off the pivot to protect the bearing when the compass is not in use.

LEVEL VIALS: Two plate level vials (fig. 1.11.3) are placed at right angles to each other. On many transits, one plate level vial is mounted on the left side, attached to the standard, under the vertical circle vernier.

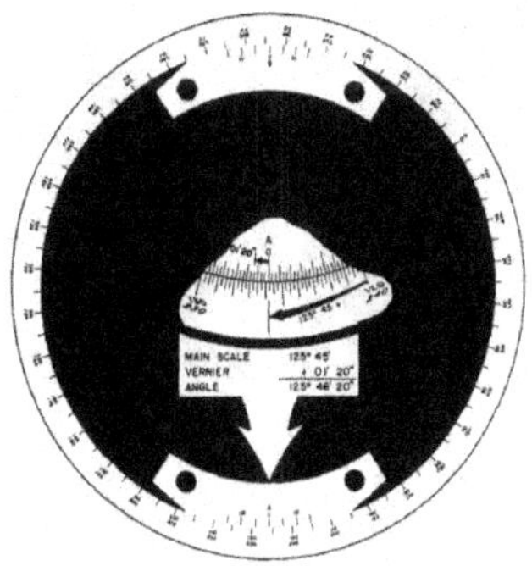

Figure 1.11.3: Horizontal Scales, 20 Second Transit

The other vial is then parallel to the axis of rotation for the vertical motion. The sensitivity of the plate level vial bubbles is about 70 sec of movement for 2 mm of tilt. Most engineer's transits have a level vial mounted on the telescope to level it. The sensitivity of this bubble is about 30 sec per 2- mm tilt.

Direct Vernier and Vernier for Circles of a unit.

CIRCLES AND VERNIERS: The horizontal and vertical circles and their verniers are the parts of the engineer's transit by which the values of horizontal and vertical angles are determined. A stadia arc is also included with the vertical circle on some transits.

The horizontal circle and verniers of the transit that are issued to SEABEE units are graduated to give least readings of either 1 min or 20 sec of arc. The horizontal circle is mounted on the lower plate. It is graduated to 15 min for the 20-sec transit (fig. 1.11.3) and 30 min for the 1-min transit (fig. 1.11.4). The plates are numbered from 0° to 360°, starting with a common point and running both ways around the circle. Two double verniers, known as the A and B verniers, are mounted on the upper plate with their indexes at circle readings 180 o apart. A double vernier is one that can be read in both directions from the index line. The verniers reduce the circle graduations to the final reading of either 20 sec or 1 min.

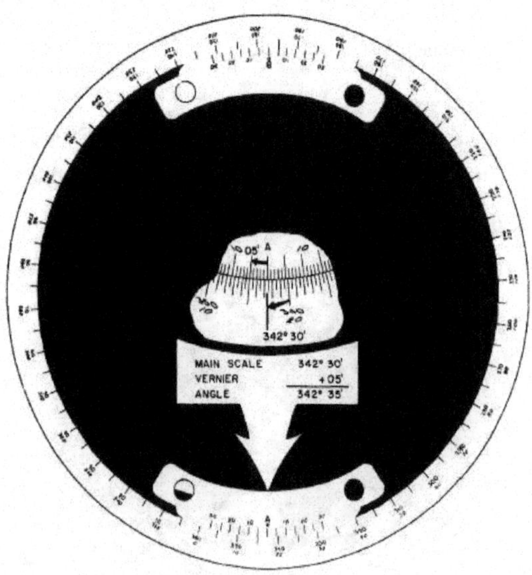

Figure 1.11.4: Horizontal Scales, 1-Minute Transit

The A vernier is used when the telescope is in its normal position, and the B vernier is used when the telescope is plunged.

The VERTICAL CIRCLE of the transit (fig. 1.11.4) is fixed to the horizontal axis so it will rotate with the telescope. The vertical circle normally is graduated to 30′ with 10° numbering. Each quadrant is numbered from 0° to 90°; the 00 graduations define a horizontal plane, and the 90° graduations lie in the vertical plane of the instrument. The double vernier used with the circle is attached to the left standard of the transit, and its least reading is 1′. The left half of the double vernier is used for reading angles of depression, and the right half of this vernier is used for reading angles of elevation. Care must be taken to read the vernier in the direction that applies to the angle observed.

In addition to the vernier, the vertical circle may have an H and V (or HOR and VERT) series of graduations, called the STADIA ARC (fig. 1.11.5). The H scale is adjusted to read 100 when the line of sight is level, and the graduations decrease in both directions from the level line. The other scale, V, is graduated with 50 at level, to 10 as the telescope is depressed, and to 90 as it is elevated.

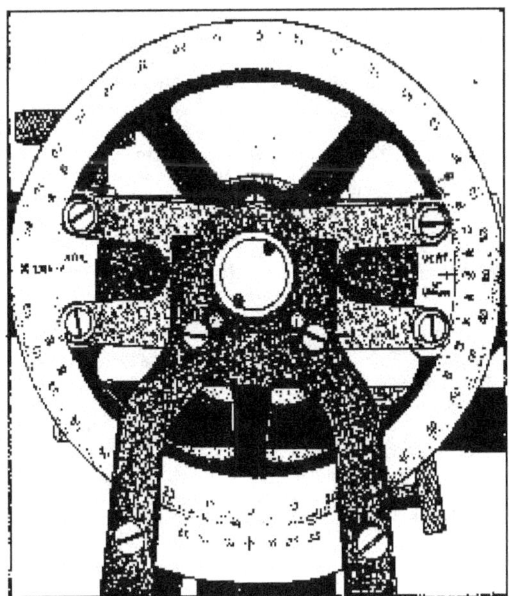

Figure 1.11.5: Vertical Circle with Verniers, Scales, and Stadia Arc.

The VERNIER, or vernier scale, is an auxiliary device by which a uniformly graduated main scale can be accurately read to a fractional part of a division. Both scales may be straight as on a leveling rod or curved as on the circles of a transit. The vernier is uniformly divided, but each division is either slightly smaller (direct vernier) or slightly larger (retrograde vernier) than a division of the main scale (fig. 1.11.5). The amount a vernier division differs from a division of the main scale determines the smallest reading of the scale that can be made with the particular vernier. This smallest reading is called the LEAST COUNT of the vernier. It is determined by dividing the value of the smallest division on the scale by the number of divisions on the vernier.

1. Direct Vernier

A scale graduated in hundredths of a unit is shown in figure fig. 1.11.5, view A, and a direct vernier for reading it to thousandths of a unit. The length of 10 divisions on the vernier is equal to the length of 9 divisions on the main scale. The index, or zero of the vernier, is set at 0.340 unit. If the vernier were moved 0.001 unit toward the 0.400 reading, the Number 1 graduation of the vernier shown in figure fig. 1.11.5, view A, would coincide with 0.35 on the scale, and the index would be at 0.341 unit. The vernier, moved to where graduation Number 7 coincides with 0.41 on the scale, is shown in figure 1.11.5, view B. In this position, the correct scale reading is 0.347 unit (0.340 + 0.007). The index with the zero can be seen to point to this reading. Retrograde Vernier. A retrograde vernier on which each division is 0.001 unit longer than the 0.01 unit divisions on the main scale is shown in figure fig. 1.11.5, view C. The length of the 10 divisions on the vernier equals the length of the 11 divisions of the scale. The retrograde vernier extends from the index, backward along the scale. Figure fig. 1.11.5, view D, shows a scale reading of 0.347 unit, as read with the retrograde vernier.

2. Vernier for Circles

Views E and F of figure 1.11.15 represent part of the horizontal circle of a transit and the direct vernier for reading the circle. The main circle graduations are numbered both clockwise and counter clockwise. A double vernier that extends to the right and to the left of the index makes it possible to read the main circle in either direction. The vernier to the left of the index is used for reading clockwise angles, and the vernier to the right of the index is used for reading.

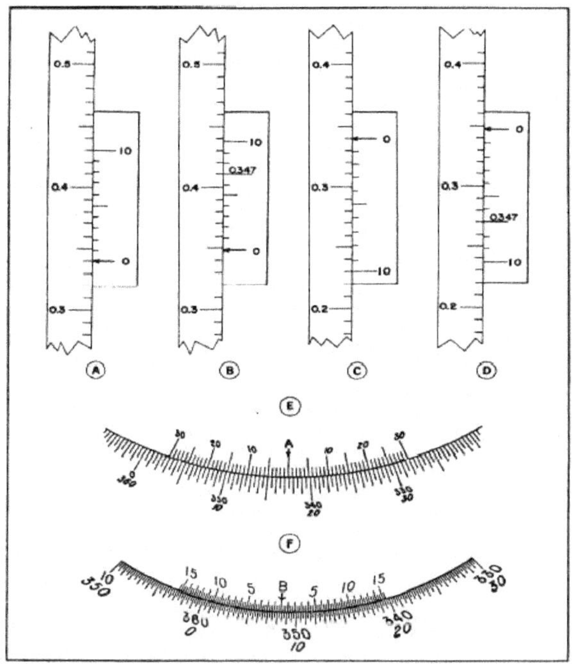

Figure 1.11.6

Counter clockwise angles. The slope of the numerals in the vernier to be used corresponds to the slope of the numerals in the circle being read. Care must be taken to use the correct vernier. In figure 1.11.6, view E, the circle is graduated to half degrees, or 30 min. On this vernier, 30 divisions are equal in length to 29 divisions on the circle. The least reading of this vernier is 30 min divided by 30 divisions, or 1 min.

3. Theodolite

A theodolite is essentially a transit of high precision. Theodolites come in different sizes and weights and from different manufacturers. Although theodolites may differ in appearance, they are basically alike in their essential parts and operation. Some of the models currently available for use in the military are WILD (Herrbrugg), BRUNSON, K&E, (Keuffel & Esser), and PATH theodolites.

To give you an idea of how a theodolite differs from a transit, we will discuss some of the most commonly used theodolites in the U.S. Armed Forces.

1.12. One-Minute Theodolite

The 1-min directional theodolite is essentially a directional type of instrument. This type of instrument can be used, however, to observe horizontal and vertical angles, as a transit does.

The theodolite shown in figure 1.12.1 is a compact, lightweight, dustproof, optical reading instrument. The scales read directly to the nearest minute or 0.2 mil and are illuminated by either natural or artificial light. The main or essential parts of this type of theodolite are discussed in the next several paragraphs.

1.12.1. Horizontal Motion

Located on the lower portion of the alidade, and adjacent to each other, are the horizontal motion clamp and tangent screw used for moving the theodolite in azimuth. Located on the horizontal circle casting is a horizontal circle clamp that fastens the circle to the alidade. When this horizontal (repeating) circle clamp is in the lever-down position, the horizontal circle turns with the telescope. With the circle clamp in the lever-up position, the circle is unclamped and the telescope turns independently. This combination permits use of the theodolite as a REPEATING INSTRUMENT. To use the theodolite as a DIRECTIONAL TYPE OF INSTRUMENT, you should use the circle clamp only to set the initial reading. You should set an initial reading of $0° \, 30°$ on the plates when a direct and reverse (D/R) pointing is required. This will minimize the possibility of ending the D/R pointing with a negative value.

1.12.2. Vertical Motion

Located on the standard opposite the vertical circle are the vertical motion clamp and tangent screw. The tangent screw is located on the lower left and at right angles to the clamp. The telescope can be rotated in the vertical plane completely around the axis (360°).

1. LEVELS: The level vials on a theodolite are the circular, the plate, the vertical circle, and the telescope level. The CIRCULAR LEVEL is located on the tribrach of the instrument and is used to roughly level the instrument. The PLATE LEVEL, located between the two standards, is used for leveling the instrument in the horizontal plane. The VERTICAL CIRCLE LEVEL (vertical collimation) vial is often referred to as a split bubble. This level vial is completely built in, adjacent to the vertical circle, and viewed through a prism and $45°$ mirror system from the eyepiece end of the telescope. This results in the viewing of one- half of each end of the bubble at the same time. Leveling consists of bringing the two halves together into exact coincidence, as shown in figure 1.12.1.

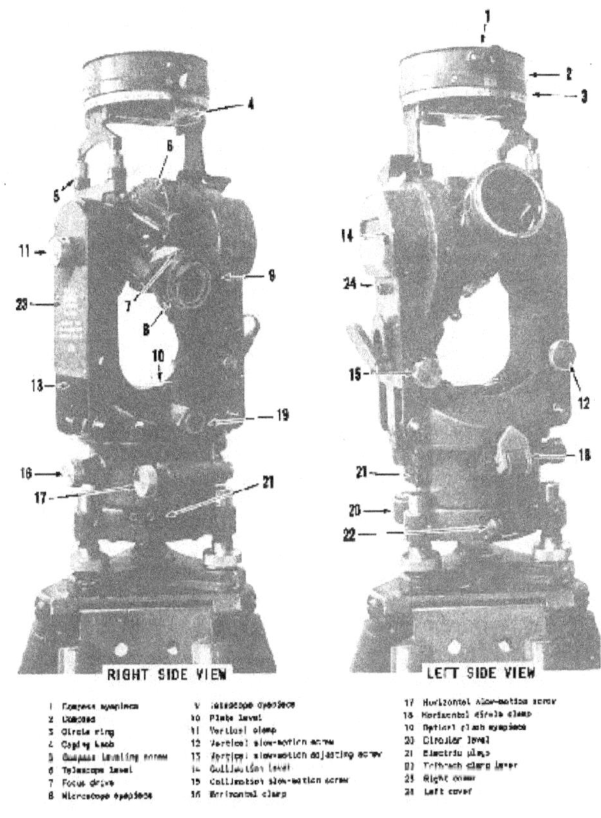

Figure 1.12.1: One-Minute Theodolite

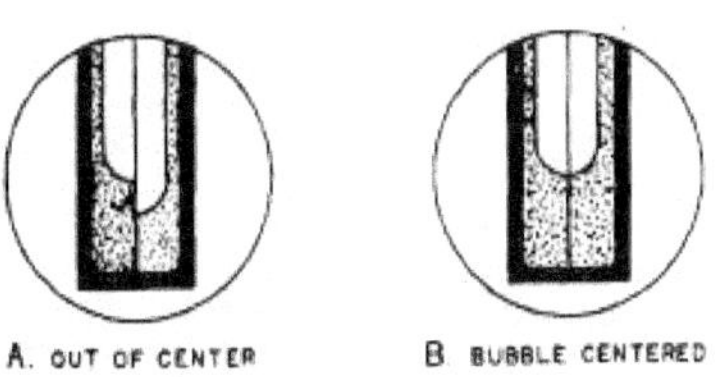

Figure 1.12.2: Coincidence-Type Level

The TELESCOPE LEVEL, mounted below the telescope, uses a prism system and a 45^0 mirror for leveling operations. When the telescope is plunged to the reverse position, the level assembly is brought to the top.

TELESCOPE: The telescope of a theodolite can be rotated around the horizontal axis for direct and reverse readings. It is a 28-power instrument with the shortest focusing distance of about 1.4 meters. The cross wires are focused by turning the eyepiece; the image, by turning the focusing ring. The reticle (fig. 1.12.3) has horizontal and vertical cross wires, a set of vertical and horizontal ticks (at a stadia ratio of 1:100), and a solar circle on the reticle for making solar observations. This circle covers 31 min of arc and can be imposed on the sun's image (32 min of arc) to make the pointing refer to the sun's center. One-half of the vertical line is split for finer centering on small distant objects.

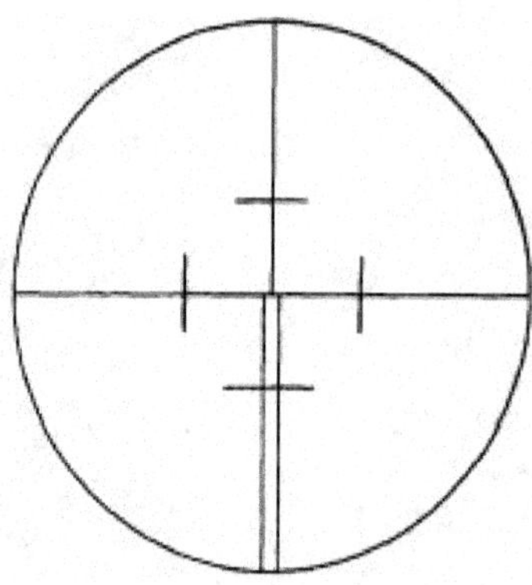

Figure 1.12.3: Theodolite Reticle

The telescope of the theodolite is an inverted image type. Its cross wires can be illuminated by either sunlight reflected by mirrors or by battery source. The amount of illumination for the telescope can be adjusted by changing the position of the illumination mirror.

TRIBRACH: The tribrach assembly (fig. 1.12.4), found on most makes and models, is a detachable part of the theodolite that contains the leveling screw, the circular level, and the optical plumbing device. A locking device holds the alidade and the tribrach together and permits interchanging of instruments without moving the tripod.

In a "leapfrog" method, the instrument (alidade) is detached after observations are completed. It is then moved to the next station and another tribrach. This procedure reduces the amount of instrument setup time by half.

CIRCLES: The theodolite circles are read through an optical microscope. The eyepiece is located to the right of the telescope in the direct position, and to the left, in the reverse.

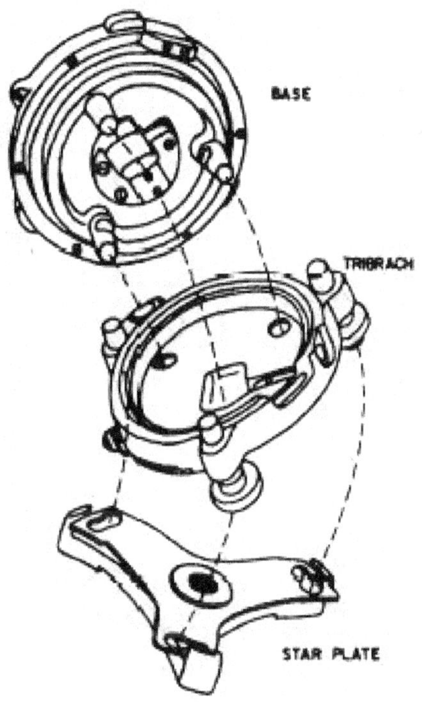

Figure 1.12.4: Three-screw Leveling Head

The microscope consists of a series of lenses and prisms that bring both the horizontal and the vertical circle images into a single field of view. In the DEGREE-GRADUATED SCALES (fig. 1.12.5), the images of both circles are shown as they would appear through the microscope of the 1-min theodolite. Both circles are graduated from 0° to 360° with an index graduation for each degree on the main scales. This scale's graduation appears to be superimposed over an auxiliary that is graduated in minutes to cover a span of 60 min (1°). The position of the degree mark on the auxiliary scale is used as an index to get a direct reading in degrees and minutes. If necessary, these scales can be interpolated to the nearest 0.2 min of arc. The vertical circle reads 0° when the theodolite's telescope is pointed at the zenith, and 180° when it is pointed straight down. A level line reads 90° in the direct position and 270^0 in the reverse. The values read from the vertical circle are referred to as ZENITH DISTANCES and not vertical angles. Figure 1.12.6 shows how these zenith distances can be converted into vertical angles.

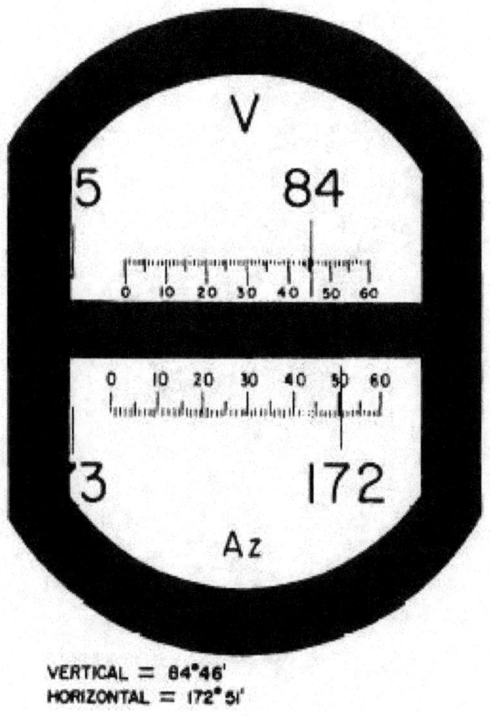

Figure 1.12.5: Degree-graduated Scales

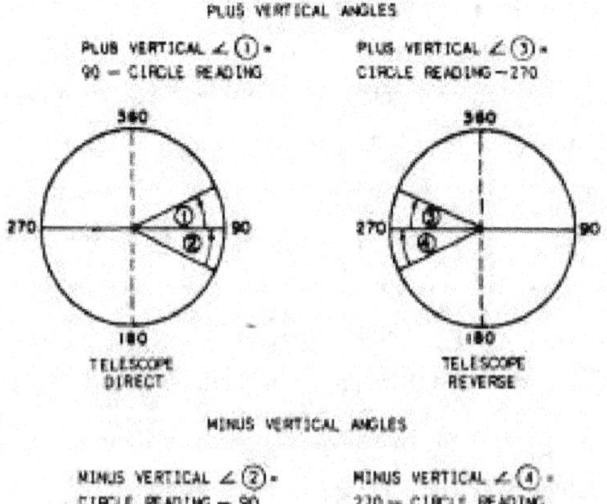

Figure 1.12.6: Converting Zenith Distances into Vertical Angles (Degrees)

In the MIL-GRADUATED SCALES (fig. 1.12.7), the images of both circles are shown as they would appear through the reading micro-scope of the 0.2-mil theodolite. Both circles are graduated from 0 to 6,400 mils. The main scales are marked and numbered every 10 mils, with the last zero dropped.

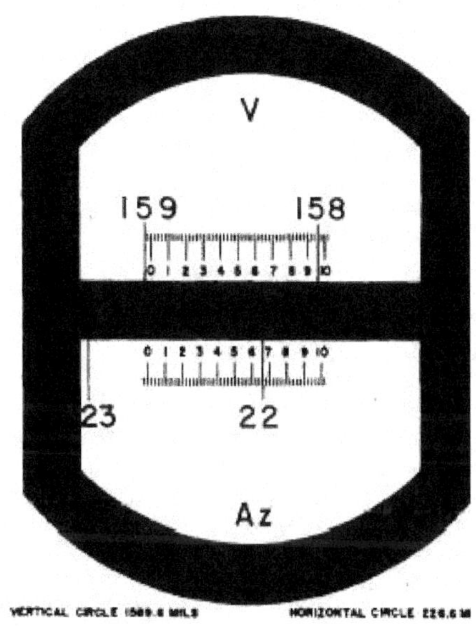

Figure 1.12.7: Mil-Graduated Scales

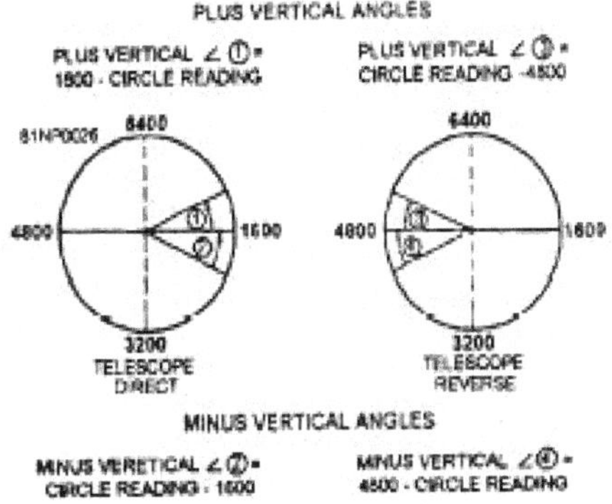

Figure 1.12.8: Vertical Angles from Zenith Distances (Mils)

The auxiliary scales are graduated from 0 to 10 roils in 0.2-mil increments. Readings on the auxiliary scale can be interpolated to 0.1 mil. The vertical circle reads 0 mil when the telescope is pointed at the zenith, and 3,200 mils when it is pointed straight down. A level line reads 1,600 roils in the direct position and 4,800 roils in the reverse. The values read are zenith distances. These zenith distances can be converted into vertical angles The excavation of material in underwater areas is called dredging, and a dredge is an excavator afloat on a barge. A dredge may get itself into position by cross bearings, taken from the dredge on objects of known location on the beach, or by some other piloting method. Many times, however, dredges are positioned by survey triangulation. The method of determining direction angles from base line control points is the same as that just described.

Land Surveying

Land surveying includes surveys for locating and monumenting the boundaries of a property; preparation of a legal description of the limits of a property and of the area included; preparation of a property map; resurveys to recover and remonument property corners; and surveys to subdivide property. It is sometimes necessary to retrace surveys of property lines, to reestablish lost or obliterated corners, and to make ties to property lines and corners; for example, a retracement survey of property lines may be required to assure that the military operation of quarry excavation does not encroach on adjacent property where excavation rights have not been obtained. Similarly, an access road from a public highway to the quarry site, if it crosses privately owned property, should be tied to the property lines that are crossed so that correctly executed easements can be obtained to cross the tracts of private property.

EAs may be required to accomplish property surveys at naval activities outside the continental limits of the United States for the construction of naval bases and the restoration of such properties to property owners. The essentials of land surveying as practiced in various countries are similar in principle. Although the principles pertaining to the surveys of public and private lands within the United States are not necessarily directly applicable to foreign countries, a knowledge of these principles will enable the EA to conduct the survey in a manner required by the property laws of the nation concerned.

In the United States, land surveying is a survey conducted for the purpose of ascertaining the correct boundaries of real estate property for legal purposes. In accordance with federal and states laws, the right and/or title to landed property in the United States can be transferred from one person to another only by means of a written document, commonly called a deed.

To constitute a valid transfer, a deed must meet a considerable number of legal requirements, some of which vary in different states. In all the states, however, a deed must contain an accurate description of the boundaries of the property.

A right in real property need not be complete, outright ownership (called fee simple). There are numerous lesser rights, such as leasehold (right to occupancy and use for a specified term) or easement (right to make certain specified use of property belonging to someone else). But in any case, a valid transfer of any type of right in real property usually involves an accurate description of the boundaries of the property.

As mentioned previously, the EA may be required to perform various land surveys. As a survey team or crew leader, you should have a knowledge of the principles of land surveys in order to plan your work accordingly.

Property Boundary Description

A parcel of land may be described by metes and bounds, by giving the coordinates of the property corners with reference to the plane coordinates system, by a deed reference to a description in a previously recorded deed, or by References to block and individual property numbers appearing on a recorded map.

By Metes and Bounds

When a tract of land is defined by giving the bearings and lengths of all boundaries, it is said to be described by metes and bounds. This is an age-old method of describing land that still forms the basis for the majority of deed descriptions in the eastern states of the United States and in many foreign lands. A good metes-and-bounds description starts at a point of beginning that should be monumented and referenced by ties or distances from well-established monuments or other reference points. The bearing and length of each side is given, in turn, around the tract to close back on the point of beginning. Bearing may be true or magnetic grid, preferably the former. When magnetic bearings are read, the declination of the needle and the date of the survey should be stated. The stakes or monuments placed at each corner should be described to aid in their recovery in the future. Ties from corner monuments to witness points (trees, poles, boulders, ledges, or other semi permanent or permanent objects) are always helpful in relocating corners, particularly where the corner markers themselves lack permanence. In timbered country, blazes on trees on or adjacent to a boundary line are most useful in re-establishing the line at a future date. It is also advisable to state the names of abutting property owners along the several sides of the tract being described.

Many metes-and-bounds descriptions fail to include all of these particulars and are frequently very difficult to retrace or locate in relation to adjoining ownerships.

One of the reasons why the determination of boundaries in the United States is often difficult is that early surveyors often confined themselves to minimal description; that is, to a bare statement of the metes. Today, good practice requires that a land surveyor include all relevant information in his description.

In preparing the description of a property, the surveyor should bear in mind that the description must clearly identify the location of the property and must give all necessary data from which the boundaries can be re-established at any future date. The written description contains the greater part of the information shown on the plan. Usually both a description and a plan are prepared and, when the property is transferred, are recorded according to the laws of the county concerned. The metes-and-bounds description of the property shown in figure 10-34 is given below.

"All that certain tract or parcel of land and premises, hereinafter particularly described, situate, lying and being in the Township of Maplewood in the County of Essex and State of New Jersey and constituting lot 2 shown on the revised map of the Taylor property in said township as filed in the Essex County Hall of Records on March 18, 1944.

"Beginning at an iron pipe in the north-westerly line of Maplewood Avenue therein distant along same line four hundred and thirty-one feet and seventy- one-hundredths of a foot north-easterly from a stone monument at the northerly corner of Beach Place and Maplewood Avenue; thence running (1) North forty-four degrees thirty-one and one-half minutes West along land of..."

Another form of a lot description maybe presented as follows:

"Beginning at the north easterly corner of the tract herein described; said corner being the intersection of the southerly line of Trenton Street and the westerly line of Ives Street; thence running S6°29'54"E bounded easterly by said Ives Street, a distance of two hundred and twenty-seven one hundredths (200.27) feet to the northerly line of Wickenden Street; thence turning an interior angle of 89°59'16" and running S83°39'50"W bonded southerly by said Wickenden Street, a distance of one hundred and no one-hundredths (100.00) feet to a corner; thence turn-ing an interior angle of...."

You will notice that in the above example, interior angles were added to the bearings of the boundary lines. This will be another help in retracing lines.

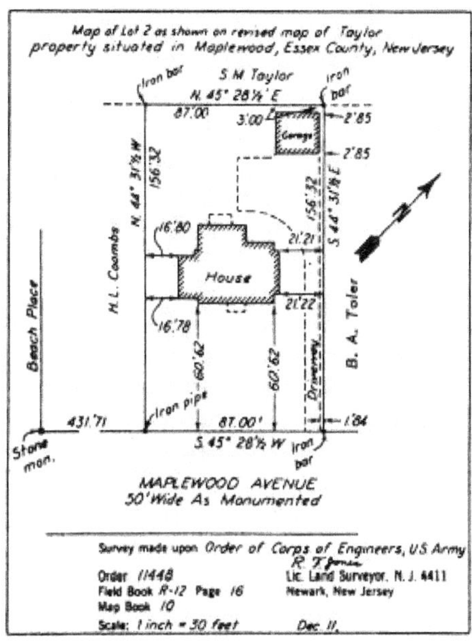

Figure 1.12.9: Lot Plan by Metes and Bounds

1.13. Intro to Antique Survey Instruments

First, some basics about their composition and finish... most instruments were made of wood, brass, or aluminum, although you will find whole instruments or instrument parts made of iron, steel, ebony, ivory, celluloid, and plastic. It is important to remember that many surveying instruments were "needle" instruments and their magnetic needles would not seek north properly if there were local sources of interference, such as iron. The United States General Land Office issued instructions requiring brass Gunters chains to be used in close proximity to the magnetic needle. (They soon changed that requirement to steel brazed link chains; the brass chain could not stand up to the type of wear and tear a chain received.

In American surveying instruments, wood was common until about 1800; brass instruments were made approximately 1775 to 1975, and aluminum instruments from 1885 to the present.

The finish of instruments has changed. Early wooden instruments were generally unfinished and were usually made of tight grained woods which resisted water well. Early brass instruments were usually unfinished or polished and lacquered to retain the shine.

In the mid-1800s American instrument makers began finishing brass instruments with dark finishes for two reasons: first, that the dark finish reduced glare and as a result reduced eyestrain, and secondly, that the dark finish helped to even out the heating of an instrument in the sunlight and as a result reduced collimation problems caused by the heating. Beware of being taken in by polished and lacquered brass instruments; prior to 1900 that may have been the original finish for the instrument, but after 1900, bright brass finishes are usually not original finishes.

There are three kinds of surveying instruments that are rather unique to North American surveying. They are the compass, the chain and the transit. In addition, the engineer's or surveyor's level contributed very strongly to making the United States the leading industrial nation in the world by virtue of the highly efficient railroad systems it helped design in the mid 1800's. I take a great deal of satisfaction in pointing out that in this country it was the compass and chain that won the west, not the six-shooter!

The following is a list of antique surveying instruments and tools with a brief and basic description of how they were used.

ABNEY HAND LEVEL - Measures vertical angles.

ALIDADE - Used on a Plane Table to measure vertical and horizontal angles & distances. ALTAAZIMUTH INSTRUMENT - Measures horizontal and vertical angles; for position "fixing".

ASTRONOMIC TRANSITS - Measures vertical angles of heavenly bodies; for determining geographic position.

BAROMETER, ANEROID - Measures elevations; used to determine vertical distance. BASE-LINE BAR - Measures horizontal distances in triangulation and trilateration surveys.

BOX SEXTANT - Measures vertical angles to heavenly bodies. CHRONOGRAPH - Measures time.

CHRONOMETER - Measures time.

CIRCUMFERENTER - Measures horizontal directions and angles. CLINOMETER - Measures vertical angles.

COLLIMATOR - For adjusting and calibrating instruments.

COMPASSES - Determines magnetic directions; there are many kinds, including plane, vernier, solar, telescopic, box, trough, wet, dry, mariners, prismatic, pocket, etc.

CROSS, SURVEYORS - For laying out 90 and 45 degree angles.

CURRENT METER - Measures rate of water flow in streams and rivers.

DIAL, MINER'S - A theodolite adapted for underground surveying; measures directions as well as horizontal and vertical angles.

GONIOMETER - Measures horizontal and vertical angles.

GRADIOMETER - Also known as Gradiometer level, it measures slight inclines and level lines-of-sight.

HELIOGRAPH - Signalling device used in triangulation surveys.

HELIOSTAT - Also known as a heliotrope, it was used to make survey points visible at long distances, particularly in triangulation surveys.

HORIZON, ARTIFICIAL - Assists in establishing a level line of sight, or "horizon". HYPSOMETER - Used to estimate elevations in mountainous areas by measuring the boiling points of liquids. This name was also given to an instrument which determined the heights of trees.

INCLINOMETER - Measures slopes and/or vertical angles.

LEVEL - Measures vertical distances (elevations). There are many kinds, including Cooke's, Cushing's, Gravatt. Dumpy, hand or pocket, wye, architect's, builder's, combination, water, engineer's, etc.

LEVELLING ROD - A tool used in conjunction with a levelling instrument. LEVELLING STAVES - Used in measuring vertical distances.

MINER'S COMPASS - Determines magnetic direction; also locates ore. MINER'S PLUMMET - A "lighted" plumb bob, used in underground surveying.

MINING SURVEY LAMP - Used in underground surveying for vertical and horizontal alignment.

OCTANT - For measuring the angular relationship between two objects. PEDOMETER - Measures paces for estimating distances. PERAMBULATOR - A wheel for measuring horizontal distances.

PHOTO-THEODOLITE - Determines horizontal and vertical positions through the use of "controlled" photographs.

PLANE TABLE - A survey drafting board for map-making with an alidade.

PLUMB BOB - For alignment; hundreds of varieties and sizes. PLUMMETS - Same as plumb bob.

QUADRANT - For measuring the angular relationship between two objects. RANGE POLES - For vertical alignment and extending straight lines. SEMICIRCUMFERENTER - Measures magnetic directions and horizontal angles.

SEXTANTS - Measures vertical angles; there are many kinds, including box, continuous arc, sounding, surveying, etc.

SIGNAL MIRRORS - For communicating over long distances; used in triangulation surveys. STADIA BOARDS - For measuring distances; also known as stadia rods.

STADIMETER or STADIOMETER - For measuring distances.

TACHEOMETER - A form of theodolite that measures horizontal and vertical angles, as well as distances.

TAPES - For measuring distances; made of many materials, including steel, invar, linen, etc. Also made in many styles, varieties, lengths, and increments.

THEODOLITE - Measures horizontal and vertical angles. Its name is one of the most misused in surveying instrument nomenclature, and is used on instruments that not only measure angles, but also directions and distances. There are many kinds, including transit, direction, optical, solar, astronomic, etc.

TRANSIT - For measuring straight lines. Like the theodolite, the transit's name is often misused in defining surveying instruments. Most transits were made to measure horizontal and vertical angles and magnetic and true directions. There are many kinds, including astronomic, solar, optical, vernier, compass, etc.

WAYWISER - A wheel for measuring distances.

1.13.1. Land Surveying

Land surveying includes surveys for locating and monumenting the boundaries of a property; preparation of a legal description of the limits of a property and of the area included; preparation of a property map; resurveys to recover and remonument property corners; and surveys to subdivide property. It is sometimes necessary to retrace surveys of property lines, to reestablish lost or obliterated corners, and to make ties to property lines and corners; for example, a retracement survey of property lines may be required to assure that the military operation of quarry excavation does not encroach on adjacent property where excavation rights have not been obtained. Similarly, an access road from a public highway to the quarry site, if it crosses privately owned property, should be tied to the property lines that are crossed so that correctly executed easements can be obtained to cross the tracts of private property.

EAs may be required to accomplish property surveys at naval activities outside the continental limits of the United States for the construction of naval bases and the restoration of such properties to property owners. The essentials of land surveying as practiced in various countries are similar in principle. Although the principles pertaining to the surveys of public and private lands within the United States are not necessarily directly applicable to foreign countries, a knowledge of these principles will enable the EA to conduct the survey in a manner required by the property laws of the nation concerned.

In the United States, land surveying is a survey conducted for the purpose of ascertaining the correct boundaries of real estate property for legal purposes. In accordance with federal and states laws, the right and/or title to landed property in the United States can be transferred from one person to another only by means of a written document, commonly called a deed.

To constitute a valid transfer, a deed must meet a considerable number of legal requirements, some of which vary in different states. In all the states, however, a deed must contain an accurate description of the boundaries of the property.

A right in real property need not be complete, outright ownership (called fee simple). There are numerous lesser rights, such as leasehold (right to occupancy and use for a specified term) or easement (right to make certain specified use of property belonging to someone else). But in any case, a valid transfer of any type of right in real property usually involves an accurate description of the boundaries of the property.

As mentioned previously, the EA may be required to perform various land surveys. As a survey team or crew leader, you should have a knowledge of the principles of land surveys in order to plan your work accordingly.

Proceed profile leveling or longitudinal sectioning in the field.

1.13.2. Profile Leveling

Profile leveling is a method of surveying that has been carried out along the central line of a track of land on which a linear engineering work is to be constructed/ laid. The operations involved in determining the elevation of ground surface at small spatial interval along a line is called profile leveling.

Stations

The line along which the profile is to be run is to be marked on the ground before taking any observation. Stakes are usually set at some regular interval which depends on the topography, accuracy required, nature of work, scale of plotting etc. It is usually taken to be10 meter. The beginning station of profile leveling is termed as 0+00. Points at multiples of 100m from this point are termed as full stations. Intermediate points are designated as pluses.

1.13.3. Procedure

In carrying out profile leveling, a level is placed at a convenient location (say I1) not necessarily along the line of observation. The instrument is to be positioned in such a way that first backsight can be taken clearly on a B.M. Then, observations are taken at regular intervals (say at 1, 2, 3, 4) along the central line and foresight to a properly selected turning point (say TP1). The instrument is then re-positioned to some other convenient location (say I2). After proper adjustment of the instrument, observations are started from TP1 and then at regular intervals (say at 5, 6 etc) terminating at another turning point, say TP2. Staff readings are also taken at salient points where marked changes in slope occur, such as that at X.

The distance as well as direction of lines are also measured.

Field Book for Reduction of Level

Field book for Reduction of Level

Pegs	Distance(m)	Direction	Staff Reading (m)			Difference in Elevation (m)		H.I (m)	R.L(m)	Remarks
			B.S	I.S	F.S	Rise	Fall			
A			3.005					108.620	105.615	B.M.
1	0+00			2.285		0.720			106.335	
2	0+10			1.560		0.725			107.060	
3	0+20			1.785			0.225		106.835	
4	0+30			2.105			0.320		106.515	
B	0+40		2.875		3.105		1.000	108.390	105.515	T.P.$_1$
5	0+50			3.465			0.590		104.925	
X	0+53.35			3.955			0.490		104.435	
6	0+60			3.120		0.835			105.270	
7	0+70			3.015		0.105			105.375	
8	0+80			2.580		0.435			105.810	
9	0+90			1.955		0.625			106.435	
C	1+00				1.465	0.490			106.925	T.P.$_2$
			5.880		4.570	3.935	2.625			

1.13.4. Uses of Contours Maps

Contours provide valuable information about the nature of terrain. This is very important for selection of sites, determination of catchment area of a drainage basin, to find intervisibility between stations etc. Some of the salient uses of contours are described below.

Nature of Ground

To visualize the nature of ground along a cross section of interest.

To Locate Route

Contour map provides useful information for locating a route at a given gradient such as highway, canal, sewer line etc.

Intervisibility between Stations

When the intervisibility between two points cannot be ascertained by inspection of the area, it can be determined using contour map.

To Determine Catchment Area or Drainage Area

The catchment area of a river is determined by using contour map. The watershed line which indicates the drainage basin of a river passes through the ridges and saddles of the terrain around the river. Thus, it is always perpendicular to the contour lines. The catchment area contained between the watershed line and the river outlet is then measured with a planimeter.

Storage Capacity of a Reservoir

The storage capacity of a reservoir is determined from contour map. The contour line indicating the full reservoir level (F.R.L) is drawn on the contour map. The area enclosed between successive contours are measured by planimeter. The volume of water between F.R.L and the river bed is finally estimated by using either Trapezoidal formula or Prismoidal formula.

1.13.5. Characteristics of Contour

The principal characteristics of contour lines which help in plotting or reading a contour map are as follows:

1. The variation of vertical distance between any two contour lines is assumed to be uniform. Contours are continuous.

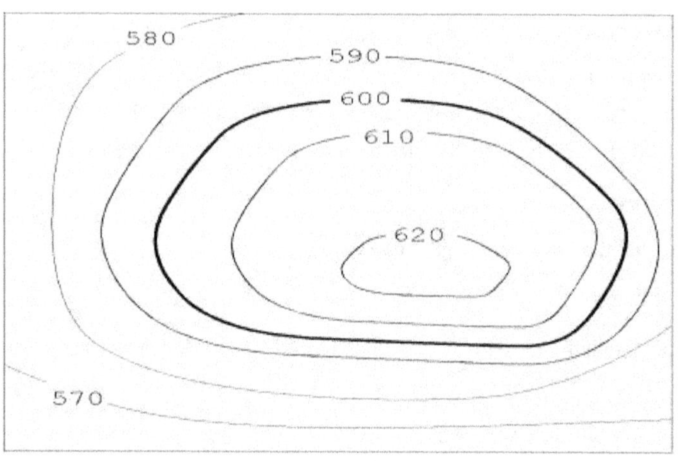

Contours are Continuous

2. The horizontal distance between any two contour lines indicates the amount of slope and varies inversely on the amount of slope. Thus, contours are spaced equally for uniform slope; closely for steep slope contours; widely for moderate slope.

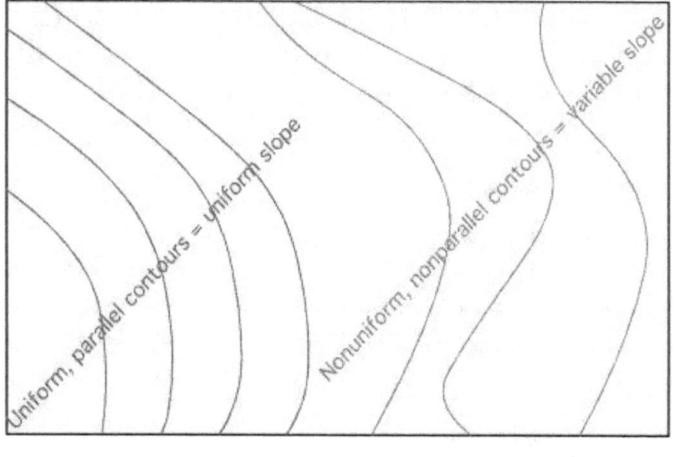

Slope

3. The steepest slope of terrain at any point on a contour is represented along the normal of the contour at that point. They are perpendicular to ridge and valley lines where they cross such lines.

4. Contours do not pass through permanent structures such as buildings.

Levelling Across a Rising Ground or Depression

While levelling across high ground, the level should not be placed on top of this high ground, but on one side so that the line of collimation just passes through the apex.

While leveling across a depression, the level should be set up on one side and not at the bottom of the depression Fig (a) and (b).

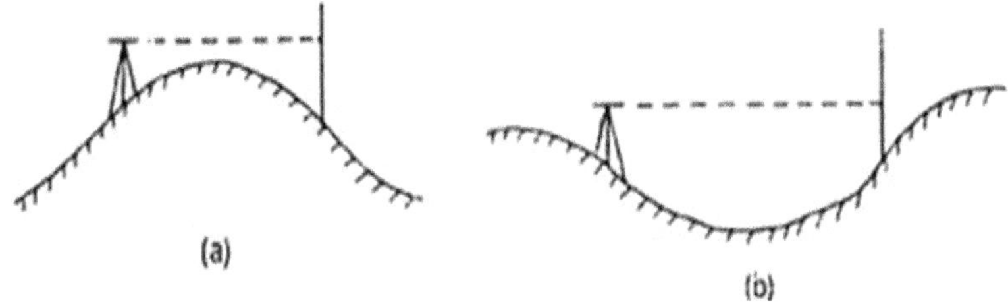

(a)

(b)

Components and their Functions of Theodolite

A compass measures the direction by measuring the angle between the line and a reference direction, which is the magnetic meridian. A compass can measure angles up to an accuracy of 30 and by judgement up to an accuracy of 15. The principle of working of the compass is based on the property of the magnetic needle, which when freely suspended, takes the north-south direction. Compass measurements are thus affected by external magnetic influences and therefore a compass is unsuitable in some areas. In this here, we will discuss another method of measuring directions of lines; a theodolite is very commonly used to measure angles in survey work.

There are a variety of theodolites-vernier, optic, electronic, etc. The improve-ments (from one form to the other) have been made to ensure ease of operation, better accuracy, and speed. Electronic theodolites display and store angles at the press of a button. This data can also be transferred to a computer for further processing. We start our discussion with the simplest theodolite-the vernier theodolite.

The vernier theodolite is a simple and inexpensive instrument but very valuable in terms of measuring angles. The common vernier theodolite measures angles up to an accuracy of 20 in a compass, where the line of sight is simple, restricting its range, theodolites are provided with telescopes which provide for much greater range and better ac-curacy in sighting distant objects. It is, however, a delicate instrument and needs to be handled carefully. The theodolite measures the horizontal angles between lines and can also measure vertical angles.

The horizontal angle measured can be the included angle, deflection angle or exterior angle in a traverse. The vertical angle is the angle in a vertical plane between the inclined line of sight of the instrument and the horizontal. In the following sections we will discuss the vernier theodolite as well as its applications in surveying.

1.14. Booking and Reducing Levels

Methods of booking and reducing the elevation of points from the observed staff readings.

(1) Collimation or Height of Instrument method.
(2) Rise and Fall method.

1.14.1. Height of Instrument Method

- In this method, the height of the instrument (Hi) is calculated for each setting of the instrument by adding back sight (plus sight) to the elevation of the B.M (First point).
- The elevation of reduced level of the turning point is then calculated by subtracting from Hi the fore sight (minus sight).
- For the next setting of the instrument, the H is obtained by adding the B.S taken on T.F. I to its R.L.
- The process continues till the R.L. of the last point (a fore sight) is obtained by subtracting the staff reading from height of the last setting of the instrument.
- If there are some intermediate points, the R.L. of those points is calculated by subtracting the intermediate sight (minus sight) from the height of the instrument for that setting.

Example

Station	B.S	I.S.	F.S.	H.I.	R.L.	Remarks
A	0.865			561.365	560.500	B.M. on Gate
B	1.025		2.105	560.285	559.260	
C		1.580			558.705	Platform
D	2.230		1.865	560.650	558.420	
E	2.355		2.835	560.270	557.815	
F			1.760		558.410	
Check	6.475		8.565		558.410	Checked
			6.475		560.500	
			2.090	Fall	2.090	

Arithmetic Check

The difference between the sum of back sights and the sum of fore sights should be equal to the difference between the last and the first R.L.

$$\Sigma \text{ B.S.} - \Sigma \text{ F.S.} = \text{Last R.L.} - \text{First R.L. RISE AND}$$

Fall Method

- In rise and fall method, the height of instrument is not at all calculated but the difference of level between consecutive points is found by comparing the staff readings on the two points for the same setting of the instrument.
- The difference between their stall readings indicates a rise or fall according the staff reading at the point is smaller or greater than that at the preceding point.
- The figures for 'rise' and 'fall' worked out thus for all the points give the vertical distance of each point above or below the preceding one, and if the level of any one point is known the level of the next will be obtained by adding its rise or subtracting its fall, as the case may be.

Example

Station	B.S.	I.S.	F.S.	Rise	Fall	R.L.	Remarks
A	0.865					560.500	B.M. on Gate
B	1.625		2.105		1.240	559.260	
C		1.580			0.555	558.705	Platform
D	2.230		1.865		0.285	558.420	
E	2.355		2.835		0.605	557.815	
F			1.760	0.595		558.410	
Check	6.475		8.565	0.595	2.685	558.410	
			6.475		0.595	560.500	Checked
		Fall	2.090	Fall	2.090	2.090	

Arithmetic Check

The difference between the sum of back sights and sum of fore sights should be equal to the difference between the sum of rise and the sum of fall and should also be equal to the difference between the R.L. of last and first point. Thus,

$$\Sigma \text{ B.S.} - \Sigma \text{ F.S.} = \Sigma \text{ RISE} - \Sigma \text{ FALL} = \text{Last R.L.} - \text{First R.L.}$$

1.15. Curvature and Refraction

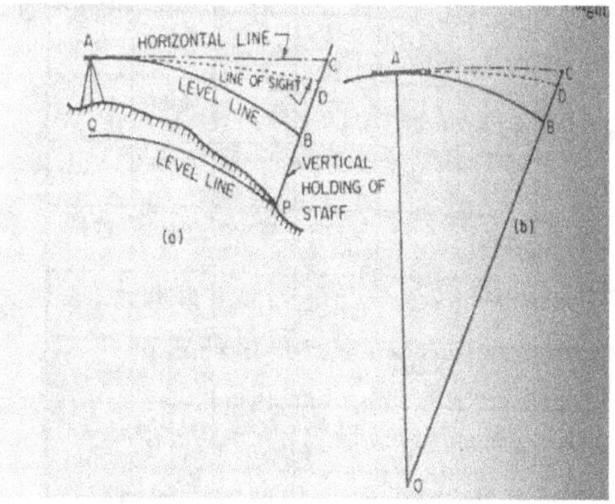

Curvature

- BC is the departure from the level line.
- Actually the staff reading should have been taken at B where the level line cuts the staff, but since the level provides only the horizontal line of sight (in the absence of refraction), the staff reading is taken at the point C.
- Thus, the apparent staff reading is more and, therefore, the object appears to be lower than it really is.
- The correction for curvature is, therefore, negative as applied to the staff reading & numerical value being equal to the amount BC.
- To find BC the value BC, we have,

$$OC^2 = OA^2 + AC^2, \ \angle CAO \text{ being } 90°$$

Let $BC = C_c = $ correction for curvature

$$AB = d = \text{horizontal distance between } A \text{ and } B$$

$$AO = R = \text{radius of earth in the same unit as that of } d$$

$$\therefore \quad (R + C_c)^2 = R^2 + d^2$$

or $\quad R^2 + 2RC_c + C_c^2 = R^2 + d^2$

$$\therefore \quad C_c (2R + C_c) = d^2$$

or $\quad C_c = \dfrac{d^2}{2R + C_c} = \dfrac{d^2}{2R}$, (Neglecting C_c in comparsion to $2R$)

Refraction

- The effect of refraction is the same as if the line of sight Was curved downward, or concave, towards earth's surface and hence the rod reading is decreased.
- Therefore, the effect of refraction is to make the objects appear high than they really are.
- The correction; as applied to staff readings is positive.
- The refraction curve is irregular because of varying atmospheric conditions, but for average conditions it is assumed to have a diameter about seven times that of the earth.

The Correction of Refraction

$$C_r = \frac{1}{7}\frac{d^2}{2R}\,(+ve) = 0.01122\,d^2 \text{ metres, when } d \text{ is in km.}$$

The combined correction due to curvature and refraction will be given by

$$C = \frac{d^2}{2R} - \frac{1}{7}\frac{d^2}{2R} = \frac{6}{7}\frac{d^2}{2R} \quad (\text{subtractive})$$

$$= 0.06735\,d^2 \text{ metres, } d \text{ being in km.}$$

The corresponding values of the corrections in English units are :

$$\left.\begin{aligned}
C_c &= \tfrac{2}{3}\,d^2 = 0.667\,d^2 \text{ feet}\\
C_r &= \tfrac{2}{21}\,d^2 = 0.095\,d^2 \text{ feet}\\
C &= \tfrac{4}{7}\,d^2 = 0.572\,d^2 \text{ feet}
\end{aligned}\right\} \quad \begin{aligned} &d \text{ is in miles and}\\ &\text{radius of earth} = 3958 \text{ miles.}\end{aligned}$$

1.16. Reciprocal Levelling

- When it is necessary to carry levelling across a river, ravine or any Obstacle requiring a long sight between two points so situated that no place for the level can be found from which the lengths of foresight and back sight will be even approximately equal, special method that is reciprocal levelling must be used.
- Let A and B be the points and observations be made with a level, the line of sight of which is inclined upwards when the Bubble is in centre of its run.
- The level is set at a point near A and staff reading are taken on A and B with the bubble in the centre of its run.
- Since B.M. A is very near to instrument, error due to curvature, refraction and collimation will be introduced in the staff readings at A.
- But there will be an error e in the staff reading on B.

- The level is then shifted to the other bank, on a point very near B.M. B, and the readings are taken on staff held at B and A.

- Since B is very near, there will be no error due to the three factors in reading the staff, but the staff reading on A will have an error e.

- Let ha and hb be the corresponding staff readings on A and B for the first set of the level and ha' and hb' be the readings for the second set.

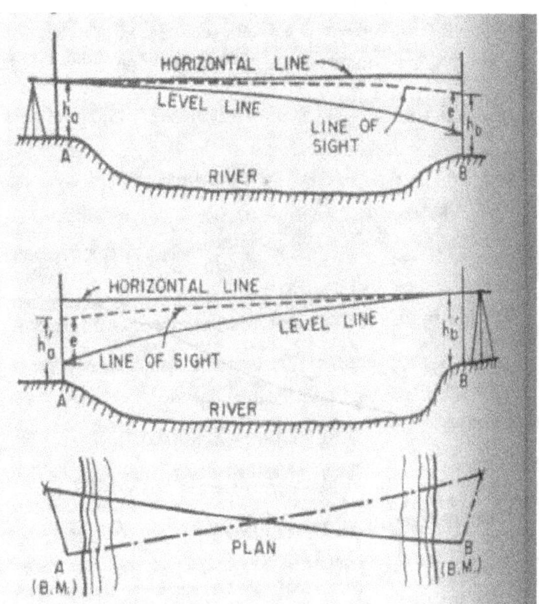

Longitudinal Sectioning

Procedure

- Profile levelling, like differential levelling, requires the establishment of turning points on which both back and foresights are taken.

- In addition, any number of intermediate sights may be obtained on points along the line from each set up of the instrument.

- It is generally best to set up the level to one side of the profile line to avoid too short sights on the points near the instrument.

- For each set up, intermediate sights should be taken after the foresight on the next turning station has been taken.

- The level is then set up in an advanced position and a back sight is taken on that turning point.

- The position of the intermediate points on the profile are simultaneously located by chaining a along the profile and noting their distances from the point of commencement.

- When the vertical profile of the ground is regular or gradually curving, levels are taken on points at equal-distances apart and generally at intervals of a chain length.

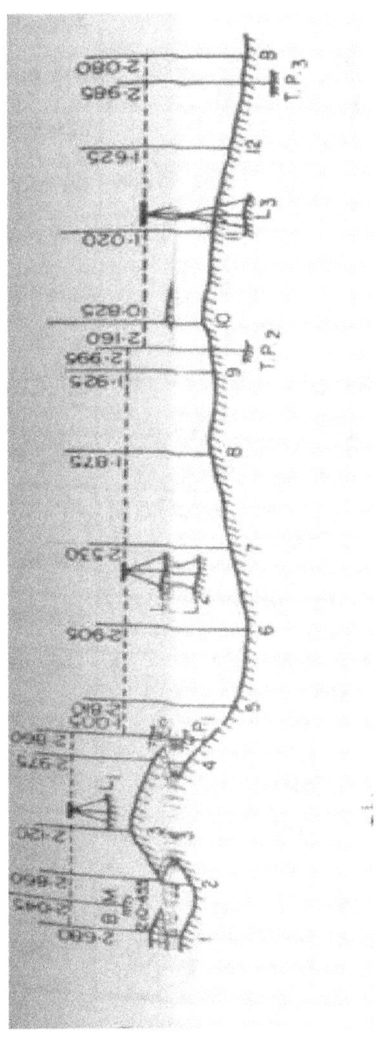

Cross Sectioning

- Cross sections are run at right angles to the longitudinal profile and o either side of it for the purpose of lateral outline of the ground Surface.
- They provide the data for estimating quantities of earth work and for other purposes.
- The cross-sections are numbered consecutively from the commencement of the centre line and are set out at right angles to the main line of section with the chain and tape, the cross-staff or the optical square and the distances are measured left and right from the centre peg.
- Cross-section be taken at each chain.
- The length of cross-section depends the nature of work.

Plotting the Profile

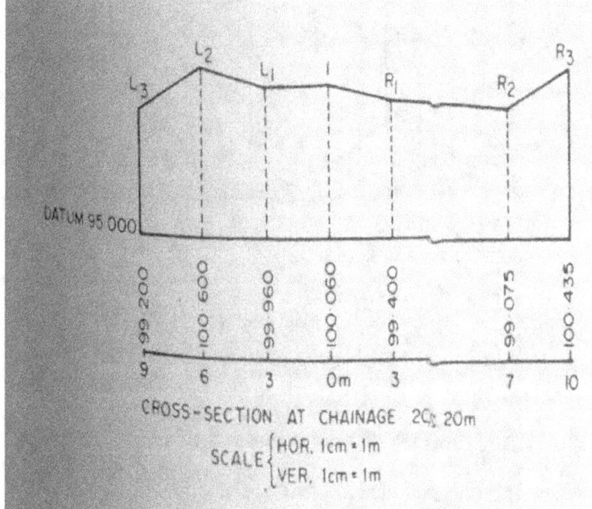

Problems

1. A steel tape of normal length 30 m was suspended between to support and measure the length of line, the measured length of slope $\theta = 3°15'$ in 29.859 m. The mean temperature during measurement was 12°C and pull applied was 100KN. If the standard length of the tape is 30.005 m at 20°C and the standard pull is 45 KN. Calculate the horizontal length. Take the weight of the tape 0.15N/m. Cross sectional area 2.15 cm², change in temperature $\alpha = 1.15 \times 10^{-5}/°C$ and $E = 2 \times 10^5$ N/mm²

Solution

Given data

$\theta = 3°15'$

Measured length = 29.859m

Tm = 12°C

Pm = 100 KN

L = 30.005 m

To = 20°C

Pa = 45KN

A = 2.5 cm²

$\alpha = 1.15 \times 10\text{-}5/°C$ and $E = 2 \times 105$ N/mm²

(i) Temperature Correction

Ct = α(Tm – To)L

 = 1.15x10-5(12-20)x 30.005

 = -0.00276m

(ii) Correction for pull

      Cp = (Pm – Po)xL/AxE

        = (100-45) x 30.005/2.5x2*105

        = 0.0033

Total correction = Ct + Cp

              = 0.00276 +0.0033

              = 0.00054 m

Corrected length = 30.005 + 0.00054

              = 30.00554 m

Actual length along the slope = L/L'x ML

                    = (30.00554/30.005)x29.859

                    = 29.86 m

Cos θ = Horizontal distance/length along slope

      = 29.86 x cos3°50'

      = 29.793 m

MODEL SHORT QUESTION AND ANSWER

UNIT I

FUNDAMENTALS OF CONVENTIONAL SURVEYING AND LEVELLING

1. **What is meant by well-conditioned triangle? (April/May 2018) (April/May 2011) (Nov/Dec 2014)**

   - A well-conditioned triangle is one in which no included angle is less than 30^0 or greater than 120^0. An equilateral triangle is the best conditioned triangle or an ideal triangle.

2. **Distinguish between survey station and tie station. (April/May 2018)**

   - A Survey Station is a point of Importance at the beginning and end of a chain line. There are two main types of stations namely Main station and Subsidiary or Tie station.

   - Any Point selected on the main survey line where it is necessary to run the auxiliary lines to locate the interior details such as fences, hedges, buildings, etc., when they are at some distance from the main survey lines are known as Subsidiary or Tie stations.

3. **What are the application of plane table surveying. (April/May 2018)**

   - It is suitable for location of details as well as contouring for large scale maps directly in the field.

   - As surveying and plotting are done simultaneously in the field, chances of getting omission of any detail get less.

   - As the instruments used are simple, not much skill for operation of instruments is required. This method of survey requires no field book.

4. **Distinguish between a check line and tie line. (Nov/Dec 2017)**

   - Base line: It is the most important line & is the longest line. Main frame works of survey line are built on it.

   - Check lines: These are the lines connecting Main station to a subsidiary station on opposite site are connecting to subsidiary station. On the sides of main lines the purpose measuring such lines is to check the accuracy within main station are located this lines are also known as group line.

   - Tie line: a tie line is a line which joins subsidiary or tie stations to the main line. The main object of running a tie line is to take the details of nearby objects but it also serves the purpose of a check line.

5. **List out the few types of obstacles in chaining. (Nov/Dec 2017)**

   The three main obstacles in chaining of a line are of the following types: 1. Chaining Free, Vision Obstructed 2. Chaining Obstructed, Vision Free 3. Chaining and Vision Both Obstructed.

6. **Define Magnetic declination. (Nov/Dec 2017) (April/May 2015)**

Magnetic Declination is defined as the horizontal angle between the true north and magnetic north at a place, at the time of observation. The magnetic needle can either be deflecting, towards east (or) west of the true meridian.

7. **Name the different types of bench marks. (Nov/Dec 2017) (April/May 2016)**
   - GTS benchmark
   - Permanent benchmark
   - Arbitrary benchmark
   - Temporary benchmark

8. **Write short notes on Ranging. (April/May 2017)**

Method of locating or establishing intermediate points on a straight line between two fixed point or two survey stations is called as ranging. There are two methods of ranging.

   - Direct Method (Two ends of survey line or stations are inter-visible).
   - Indirect Method (Two ends of survey line or stations are not inter-visible).

9. **Define true bearing and magnetic bearing. (April/May 2017)**

   - Magnetic bearing: using magnetic compass readings to navigate from beginning to end points on the Earth's surface. Magnetic compass readings show the user where the Magnetic Poles are in relation to their own location at a certain time and show that using minutes and seconds of arc.
   - True bearing refers to the geometric lines which also define loci on the surface of the earth and are thought of as "true" because they never move like the Magnetic Poles and because the North and South Geographical poles are precisely opposite "points" on the spherical Earth.

10. **What are the principal of surveying? (Nov/Dec 2016) (April/May 2017)**

The fundamental principles upon which the surveying is being carried out are:

   - Working from whole to part.
   - After deciding the position of any point, its reference must be kept from at least two permanent objects or stations whose position have already been well defined.

11. **What are the source of local attraction? (Nov/Dec 2016)**

The sources of local attraction may be natural or artificial. Natural sources include Iron ores or magnetic rocks while as artificial sources consist of steel structures, iron pipes, Current carrying conductors. The iron made surveying instruments such as metric chains, ranging rods and arrows should also be kept at a safe distance apart from compass.

12. **What are the different methods of surveying based on instruments? (Nov/Dec 2016)**

Based on various types of instruments used, surveying can be classified into six types.

- Chain surveying
- Compass surveying
- Plane table surveying
- Theodolite surveying
- Tacheometric surveying
- Photographic surveying

13. **What is the object of Surveying?**

Following are the various purposes of the surveying methods.

1. To check out the alignment of various engineering structures.
2. To calculate the areas and volumes, involved in the various engineering projects.
3. To prepare the plans and maps, sections (or) profiles, contours, etc.

14. **Name the different ways of Classification of surveying**

Classification of surveys based on,

1. Purpose of surveying
2. Nature of the field
3. Methods employed
4. Instruments used

15. **What are the instruments used for the chain surveying?**

Following instruments are used in chain surveying.

1. Chain
2. Tape
3. Ranging Rods
4. Offset Rods
5. Plumb Bob
6. Pegs
7. Cross-Staff
8. Optical Square
9. Arrows
10. Whites etc.

16. **Write the equation for correction of temperature**

Temperature Correction, $Ct = \alpha\ (Tm \sim T0)$ L Where,

α – Coefficient of thermal expansion

Tm – Mean Temperature during measurements

To – Normal Temperature at standardization

L – Measured length of the line

17. **What are arrows?**

Chain pins (or) Arrows are the steel wire of 4 mm diameter and its length may vary from 25 cm to 50 cm.

One end of the arrow is bent into a loop of a circle of 50 cm diameter and the other end is made sharp Point. Arrows are used to indicate (or to mark) the end of a chain line.

18. **What is Plumb Bob?**

Plumb bobs are used to test the verticality of ranging rods and levelling staves. It is also used to transfer the end points of the chain onto ground while measuring the distances in a hilly terrain.

19. **Define compass surveying?**

The branch of surveying in which direction of survey line are determine by a compass and their length by a chain or tape is called compass surveying. This type of survey can be used to measure large areas with reasonable speed and accuracy.

20. **What is a prismatic compass?**

Prismatic compass is an instrument used to measure the bearing of a line. It consists of a magnetic needle pivoted at the centre and is free to rotate. The area below the magnetic needle is graduated between 0 to 360 degrees.

The instrument cover consists of a sighting vane and vertical hair to align the compass along the instrument station and the staff station.

21. **Distinguish between angle and bearing?**

An angle is defined as the deviation of one straight line with respect to the other one.

Bearing is defined as the angle (or) inclination of a survey line with respect to the north-south direction.

22. **Define true meridian.**

True meridian (or) Geographical meridian is defined as the line joining the geographical north and south poles. True meridian at various places are not parallel to each other.

23.   What is magnetic meridian?

Magnetic Meridian is defined as the longitudinal axis, indicated by the freely suspended, properly balanced magnetic needle. It does not coincide with the true meridian except in certain places during the year.

24.   What is Reciprocal Levelling?

It is the method of levelling and it is used when the instrument is placed equidistant from the back staff and foreword staff stations, the difference in elevation of two stations, is equal to the difference of staff readings.

25.   Name the sources of errors in leveling?

Errors in leveling may be categorized into,

1.   Personal error

2.   Errors due to natural factor sand

3.   Instrumental error.

26.   Define levelling. What are the use of levelling?

It is a branch of surveying, the object of which is,

- To find the elevation of given or assumed datum.
- To establish points at a given elevation or at different elevation with respect to a given or assumed datum.

27.   What are the different types of levelling staff?

Target staff, Self-reading staff, and Solid staff, folding staff and Telescopic staff.

28.   What is Dumpy Level?

This is the simplest type of the levelling instrument and it is compact and stable. It consists of a telescope, rigidly fixed to the supports. It can neither be rotated about its longitudinal axis nor can it be removed from its supports. A long bubble tube is attached to the top of the telescope.

29.   What is meant by height of collimation?

The R.L. (or) elevation of the line of collimation, when the instrument is perfectly levelled, is called the Height of the Instrument.

30.   Compare height of collimation and rise and fall method

S.NO	HEIGHT OF COLLIMATION METHOD	RISE AND FALL METHOD
1.	It is more rapid and less tedious and simpler as it involving few calculation	It is more laborious and tedious involving several calculation
2.	There is no check on the RL of the intermediate points	There is check on the RL of the intermediate points
3.	Errors in intermediate RL's cannot be detected	Errors in intermediate RL's can be detected

31.   Differentiate plane and geodetic surveying. (Nov/Dec 2014)

S.NO	PLANE SURVEYING	GEODETIC SURVEYING
1.	Curvature of earth is not taken into account.	Curvature of earth is taken into account.
2.	The line joining any two points is treated as a straight line.	The line joining any two points is treated as the arc of the circle.
3.	Length up to 12KM is treated as plane surveying.	Length more than 12KM is treated as geodetic surveying.
4.	It is less accurate and less correct.	It is more accurate and more correct.

32.   Define dip and declination (Nov/Dec 2014)

The angle between the magnetic meridian and geographic meridian at a place is called Magnetic Declination at that place.

It is the angle between the direction of total magnetic field of earth and a horizontal line in magnetic meridian.

33.   Define dip and declination (Apr/May 2019)

Datum is any surface to which elevation are referred. The mean sea level affords a convenient datum world over, and elevation is commonly given so much above or below sea level.

34.   Define an agonic and isogonic line (Nov/Dec 2014)

Agonic line is the line made up of points having a zero declination.

Isogonic line is the line drawn through the points of same declination.

MODEL DETAIL QUESTIONS

1. A chain line AB comes across a pond. Two points P and R are selected on the chain line on either side of the pond. A line PQ of 300.

2. m length is set on the one side of the pond and another line PS of length 500m is run on the opposite side of PQ. It is so aligned the points Q, R and S are on the same straight line. Calculate the approximate length of the pond, if QR, RS are measured as 150m and 250m respectively. **(April/May 2018)**

3. What are the different sources of error in chain surveying? Distinguish between cumulative and compensating errors. **(April/May 2018) (April/May 2017)**

4. The following are the bearings of a closed traverse. Find which station is free from local attraction and work out the bearings. **(April/May 2018) (April/May 2017) (Nv/Dec 2016)**

SIDE	F.B	B.B
AB	292^{0}15'	111^{0}45'
BC	221^{0}45'	41^0 45"
CD	90^{0}05'	270^{0}00'
DE	80^{0}35'	261^{0}40'
EA	37^{0}00'	216^{0}30'

5. The following consecutive readings were taken with a level and 4m levelling staff as 1.904,2.653,3.906,4.026,1.964,1.702,1.592,1.261,2.542,2.006,3.145. The instrument was shifted after 4th and 7th readings. The first reading was taken on the staff held on BM of RL 100.00m. Makeup level book page, apply usual check and calculate the reduced levels of points. **(April/May 2018)**

6. A closed traverse with sides is almost that of a regular pentagon. One line of the pentagon has a bearing of 54^0 30". Compute the bearing of the remaining sides taking the side in a clockwise order. **(Nov/Dec 2017)**

7. Explain the classification of surveying. **(April/May 2017)**

8. Following consecutive staffs reading were taken with a level along a sloping ground line AB at a regular distance of 20m by using 4m levelling staff 0.352, 0.787, 1.832, 2.956, 3.758, 0.953, 1.766, 2.738, 3.872, 0.812, 2.325 and 3.137. Rule out a page of level field book, enter the above readings. RL of point A is 320.288 m. Calculate RL of all the points by rise and fall method, and workout the gradient of line AB. **(April/May 2017) (Nov/Dec 2016)**

9. What is different between surveyor compass and prismatic compass? **(Nov/Dec 2016)**

10. A compass survey was carried out around a closed traverse ABCD and the following readings were obtained:

LINE	F.B	B.B
AB	$74^0 30'$	$256^0 10'$
BC	$107^0 30'$	$286^0 30'$
CD	$225^0 10'$	$45^0 10'$
DA	$306^0 50'$	$126^0 10'$

Identify the stations affected by the local attraction and work out the corrected bearings of the lines. **(Nov/Dec 2016) (April/May 2016)**

11. What is the difference between temporary and permanent adjustments of a dumpy level? Explain the temporary adjustments made in a levelling instrument. **(April/May 2016)**

12. Explain the effects of curvature and refraction in levelling and their corrections. **(April/May 2017)**

13. What are the different sources of error in levelling and explain them in detail? **(Nov/Dec 2015)**

UNIT II

2. THEODOLITE AND TACHEOMETRIC SURVEYING

Horizontal and vertical angle measurements - Temporary and permanent adjustments - Heights and distances - Tacheometer - Stadia Constants - Analytic Lens - Tangential and Stadia Tacheometry surveying - Contour - Contouring - Characteristics of contours - Methods of contouring - Tacheometric contouring - Contour gradient - Uses of contour plan and map

2. THEODOLITE

2.1. Horizontal Motion, Vertical Motion

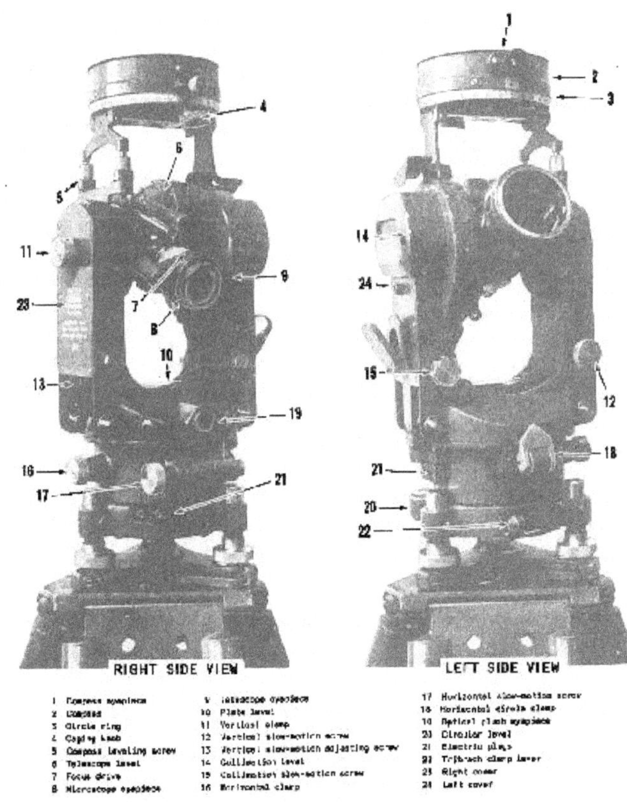

Figure 2.1

Theodolite

A theodolite is essentially a transit of high precision. Theodolites come in different sizes and weights and from different manufacturers. Although theodolites may differ in appearance, they are basically alike in their essential parts and operation. Some of the models currently available for use in the military are WILD (Herrbrugg), BRUNSON, K&E, (Keuffel & Esser), and PATH theodolites.

To give you an idea of how a theodolite differs from a transit, we will discuss some of the most commonly used theodolites in the U.S. Armed Forces.

One-Minute Theodolite

The 1-min directional theodolite is essentially a directional type of instrument. This type of instrument can be used, however, to observe horizontal and vertical angles, as a transit does.

The theodolite shown in figure 2.1.1 is a compact, lightweight, dustproof, optical reading instrument. The scales read directly to the nearest minute or 0.2 mil and are illuminated by either natural or artificial light. The main or essential parts of this type of theodolite are discussed in the next several paragraphs.

Horizontal Motion

Located on the lower portion of the alidade, and adjacent to each other, are the horizontal motion clamp and tangent screw used for moving the theodolite in azimuth. Located on the horizontal circle casting is a horizontal circle clamp that fastens the circle to the alidade. When this horizontal (repeating) circle clamp is in the lever-down position, the horizontal circle turns with the telescope. With the circle clamp in the lever-up position, the circle is unclamped and the telescope turns independently. This combination permits use of the theodolite as a REPEATING INSTRUMENT. To use the theodolite as a DIRECTIONAL TYPE OF INSTRUMENT, you should use the circle clamp only to set the initial reading. You should set an initial reading of 0° 30° on the plates when a direct and reverse (D/R) pointing is required. This will minimize the possibility of ending the D/R pointing with a negative value.

Vertical Motion

Located on the standard opposite the vertical circle are the vertical motion clamp and tangent screw. The tangent screw is located on the lower left and at right angles to the clamp. The telescope can be rotated in the vertical plane completely around the axis (360°).

LEVELS: The level vials on a theodolite are the circular, the plate, the vertical circle, and the telescope level. The CIRCULAR LEVEL is located on the tribrach of the instrument and is used to roughly level the instrument. The PLATE LEVEL, located between the two standards, is used for leveling the instrument in the horizontal plane. The VERTICAL CIRCLE LEVEL (vertical collimation) vial is often referred to as a split bubble. This level vial is completely built in, adjacent to the vertical circle, and viewed through a prism and 450 mirror system from the eyepiece end of the telescope. This results in the viewing of one- half of each end of the bubble at the same time. Leveling consists of bringing the two halves together into exact coincidence, as shown in figure 2.1.2. The TELESCOPE LEVEL, mounted below the telescope, uses a prism system and a 450 mirror for levelling operations. When the telescope is plunged to the reverse position, the level assembly is brought to the top.

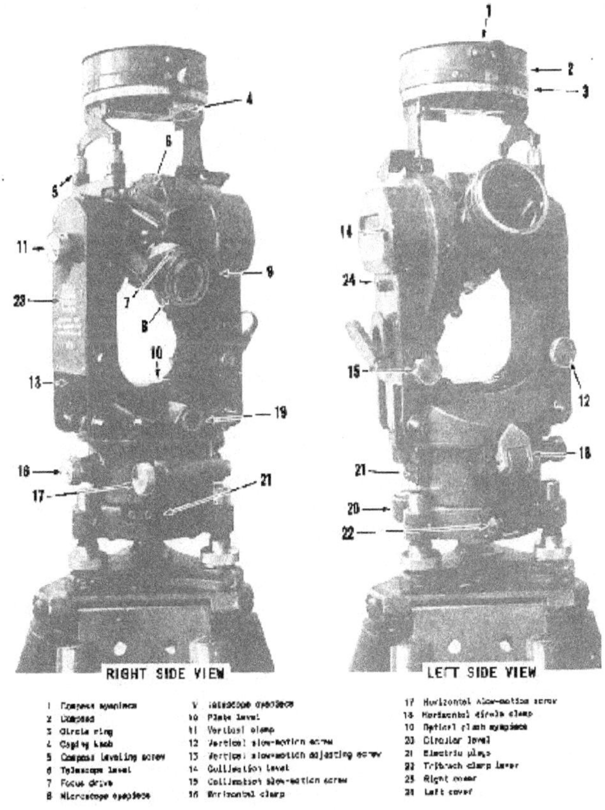

1 Compass eyepiece	9 Telescope eyepiece	17 Horizontal slow-motion screw
2 Compass	10 Plate level	18 Horizontal circle clamp
3 Circle ring	11 Vertical clamp	19 Optical plumb eyepiece
4 Capping knob	12 Vertical slow-motion screw	20 Circular level
5 Compass leveling screw	13 Vertical slow-motion adjusting screw	21 Electric plug
6 Telescope level	14 Collimation level	22 Tribrach clamp lever
7 Focus drive	15 Collimation slow-motion screw	23 Right cover
8 Microscope eyepiece	16 Horizontal clamp	24 Left cover

Figure 2.1.1: One-Minute Theodolite

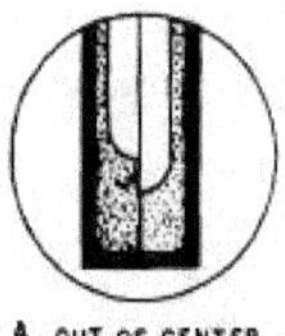
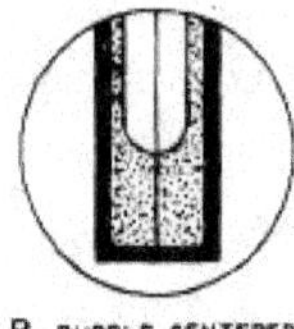

Figure 2.1.2: Coincidence-type Level

TELESCOPE: The telescope of a theodolite can be rotated around the horizontal axis for direct and reverse readings. It is a 28-power instrument with the shortest focusing distance of about 1.4 meters. The cross wires are focused by turning the eyepiece; the image, by turning the focusing ring. The reticle (fig. 2.1.3) has horizontal and vertical cross wires, a set of vertical and horizontal ticks (at a stadia ratio of 1:100), and a solar circle on the reticle for making solar observations. This circle covers 31 min of arc and can be imposed on the sun's image (32 min of arc) to make the pointing refer to the sun's center. One-half of the vertical line is split for finer centering on small distant objects.

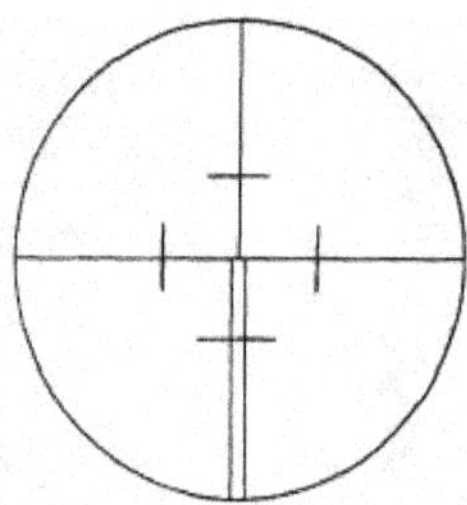

Figure 2.1.3: Theodolite Reticle

The telescope of the theodolite is an inverted image type. Its cross wires can be illuminated by either sunlight reflected by mirrors or by battery source. The amount of illumination for the telescope can be adjusted by changing the position of the illumination mirror.

TRIBRACH: The tribrach assembly (fig. 2.1.4), found on most makes and models, is a detachable part of the theodolite that contains the leveling screw, the circular level, and the optical plumbing device. A locking device holds the alidade and the tribrach together and permits interchanging of instruments without moving the tripod. In a "leapfrog" method, the instrument (alidade) is detached after observations are completed. It is then moved to the next station and another tribrach. This procedure reduces the amount of instrument setup time by half.

CIRCLES: The theodolite circles are read through an optical microscope. The eyepiece is located to the right of the telescope in the direct position, and to the left, in the reverse. The microscope consists of a series of lenses and prisms that bring both the horizontal and the vertical circle images into a single field of view.

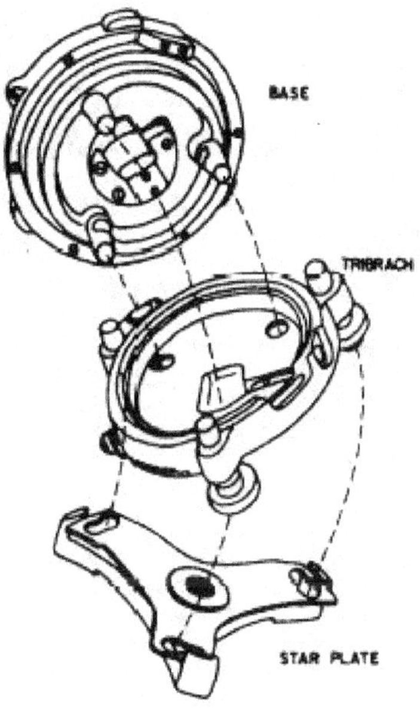

Figure 2.1.4: Three-screw Leveling Head

In the DEGREE-GRADUATED SCALES (fig. 2.1.5), the images of both circles are shown as they would appear through the microscope of the 1-min theodolite. Both circles are graduated from 0° to 360° with an index graduation for each degree on the main scales. This scale's graduation appears to be superimposed over an auxiliary that is graduated in minutes to cover a span of 60 min (1°). The position of the degree mark on the auxiliary scale is used as an index to get a direct reading in degrees and minutes. If necessary, these scales can be interpolated to the nearest 0.2 min of arc. The vertical circle reads 0° when the theodolite's telescope is pointed at the zenith, and 180° when it is pointed straight down. A level line reads 90° in the direct position and 2700 in the reverse. The values read from the vertical circle are referred to as ZENITH DISTANCES and not vertical angles. Figure 2.1.6 shows how these zenith distances can be converted into vertical angles.

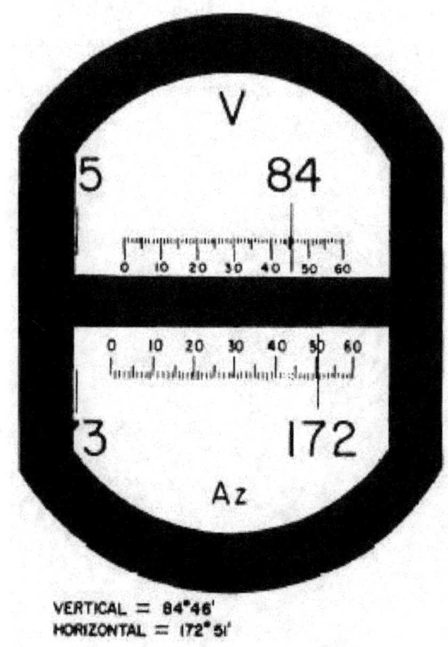

Figure 2.1.5: Degree-graduated Scales

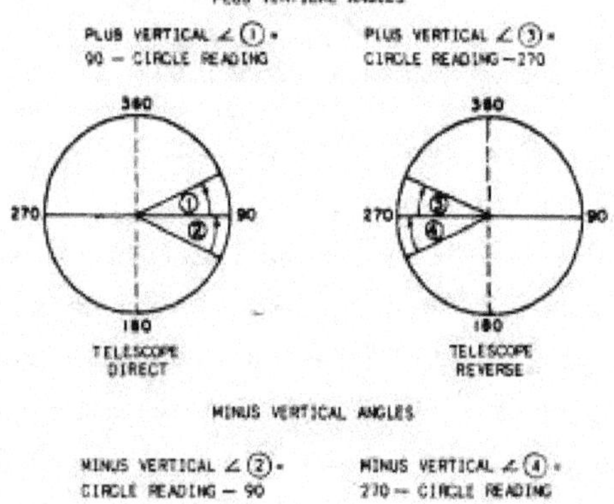

Figure 2.1.6: Converting Zenith Distances into Vertical Angles (Degrees)

In the MIL-GRADUATED SCALES (fig. 2.1.7), the images of both circles are shown as they would appear through the reading micro-scope of the 0.2-mil theodolite. Both circles are graduated from 0 to 6,400 mils. The main scales are marked and numbered every 10 mils.

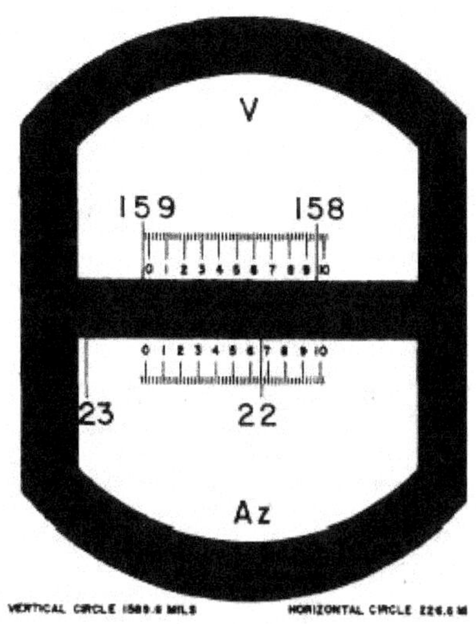

Figure 2.1.7: Mil-graduated Scales

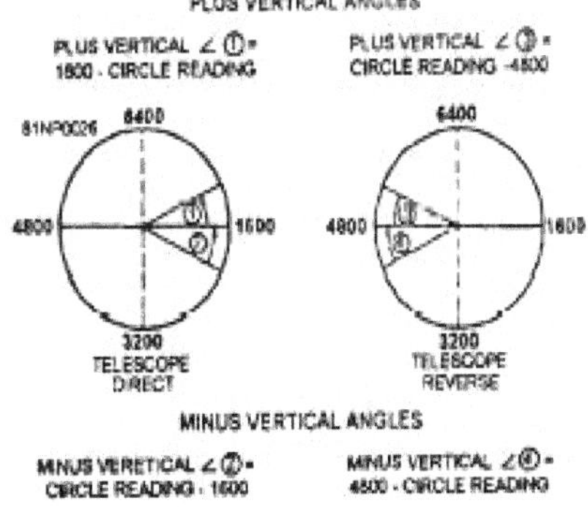

Figure 2.1.8: Vertical Angles from Zenith Distances (mils)

2.2. Property Boundary Description

A parcel of land may be described by metes and bounds, by giving the coordinates of the property corners with reference to the plane coordinates system, by a deed reference to a description in a previously recorded deed, or by References to block and individual property numbers appearing on a recorded map.

By Metes and Bounds

When a tract of land is defined by giving the bearings and lengths of all boundaries, it is said to be described by metes and bounds. This is an age-old method of describing land that still forms the basis for the majority of deed descriptions in the eastern states of the United States and in many foreign lands. A good metes-and-bounds description starts at a point of beginning that should be monumented and referenced by ties or distances from well-established monuments or other reference points. The bearing and length of each side is given, in turn, around the tract to close back on the point of beginning. Bearing may be true or magnetic grid, preferably the former. When magnetic bearings are read, the declination of the needle and the date of the survey should be stated. The stakes or monuments placed at each corner should be described to aid in their recovery in the future. Ties from corner monuments to witness points (trees, poles, boulders, ledges, or other semi-permanent or permanent objects) are always helpful in relocating corners, particularly where the corner markers themselves lack permanence. In timbered country, blazes on trees on or adjacent to a boundary line are most useful in re-establishing the line at a future date. It is also advisable to state the names of abutting property owners along the several sides of the tract being described. Many metes-and-bounds descriptions fail to include all of these particulars and are frequently very difficult to retrace or locate in relation to adjoining ownerships.

One of the reasons why the determination of boundaries in the United States is often difficult is that early surveyors often confined themselves to minimal description; that is, to a bare statement of the metes. Today, good practice requires that a land surveyor include all relevant information in his description.

In preparing the description of a property, the surveyor should bear in mind that the description must clearly identify the location of the property and must give all necessary data from which the boundaries can be re-established at any future date. The written description contains the greater part of the information shown on the plan. Usually both a description and a plan are prepared and, when the property is transferred, are recorded according to the laws of the county concerned.

"All that certain tract or parcel of land and premises, hereinafter particularly described, situate, lying and being in the Township of Maplewood in the County of Essex and State of New Jersey and constituting lot 2 shown on the revised map of the Taylor property in said township as filed in the Essex County Hall of Records on March 18, 1944.

"Beginning at an iron pipe in the north-westerly line of Maplewood Avenue therein distant along same line four hundred and thirty-one feet and seventy- one-hundredths of a foot north-easterly from a stone monument at the northerly corner of Beach Place and Maplewood Avenue; thence running (1) North forty-four degrees thirty-one and one-half minutes West along land of..."

Another form of a lot description maybe presented as follows:

"Beginning at the north-easterly corner of the tract herein described; said corner being the intersection of the southerly line of Trenton Street and the westerly line of Ives Street; thence running S6°29'54"E bounded easterly by said Ives Street, a distance of two hundred and twenty-seven one hundredths (200.27) feet to the northerly line of Wickenden Street; thence turning an interior angle of 89°59'16" and running S83°39'50"W bonded southerly by said Wickenden Street, a distance of one hundred and no one-hundredths (100.00) feet to a corner; thence turn-ing an interior angle of...."

You will notice that in the above example, interior angles were added to the bearings of the boundary lines. This will be another help in retracing lines.

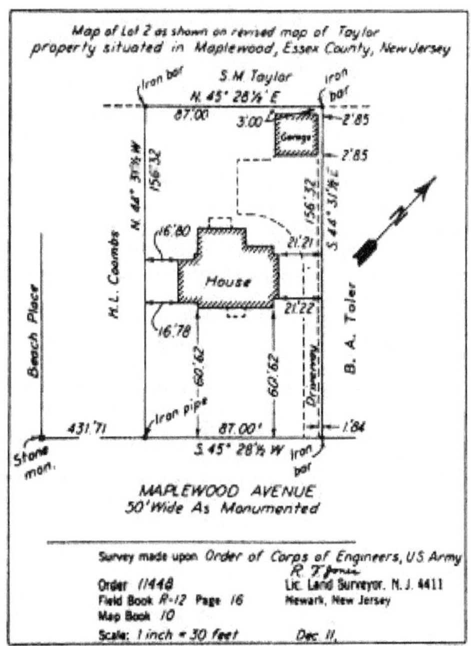

Intro to Antique Survey Instruments

First, some basics about their composition and finish... most instruments were made of wood, brass, or aluminum, although you will find whole instruments or instrument parts made of iron, steel, ebony, ivory, celluloid, and plastic. It is important to remember that many surveying instruments were "needle" instruments and their magnetic needles would not seek north properly if there were local sources of interference, such as iron. The United States General Land Office issued instructions requiring brass Gunters chains to be used in close proximity to the magnetic needle. (They soon changed that requirement to steel brazed link chains; the brass chain could not stand up to the type of wear and tear a chain received.

In American surveying instruments, wood was common until about 1800; brass instruments were made approximately 1775 to 1975, and aluminum instruments from 1885 to the present.

The finish of instruments has changed. Early wooden instruments were generally unfinished and were usually made of tight grained woods which resisted water well. Early brass instruments were usually unfinished or polished and lacquered to retain the shine. In the mid-1800s American instrument makers began finishing brass instruments with dark finishes for two reasons: first, that the dark finish reduced glare and as a result reduced eyestrain, and secondly, that the dark finish helped to even out the heating of an instrument in the sunlight and as a result reduced collimation problems caused by the heating. Beware of being taken in by polished and lacquered brass instruments; prior to 1900 that may have been the original finish for the instrument, but after 1900, bright brass finishes are usually not original finishes.

There are three kinds of surveying instruments that are rather unique to North American surveying. They are the compass, the chain and the transit. In addition, the engineer's or surveyor's level contributed very strongly to making the United States the leading industrial nation in the world by virtue of the highly efficient railroad systems it helped design in the mid 1800's. I take a great deal of satisfaction in pointing out that in this country it was the compass and chain that won the west, not the six-shooter!

The following is a list of antique surveying instruments and tools with a brief and basic description of how they were used.

ABNEY HAND LEVEL - Measures vertical angles.

ALIDADE - Used on a Plane Table to measure vertical and horizontal angles & distances. ALTAAZIMUTH INSTRUMENT - Measures horizontal and vertical angles; for position "fixing".

ASTRONOMIC TRANSITS - Measures vertical angles of heavenly bodies; for determining geographic position.

BAROMETER, ANEROID - Measures elevations; used to determine vertical distance.

BASE-LINE BAR - Measures horizontal distances in triangulation and trilateration surveys.

BOX SEXTANT - Measures vertical angles to heavenly bodies. CHRONOGRAPH - Measures time.

CHRONOMETER - Measures time.

CIRCUMFERENTER - Measures horizontal directions and angles. CLINOMETER - Measures vertical angles.

COLLIMATOR - For adjusting and calibrating instruments.

COMPASSES - Determines magnetic directions; there are many kinds, including plane, vernier, solar, telescopic, box, trough, wet, dry, mariners, prismatic, pocket, etc.

CROSS, SURVEYORS - For laying out 90 and 45 degree angles. CURRENT METER - Measures rate of water flow in streams and rivers.

DIAL, MINER'S - A theodolite adapted for underground surveying; measures directions as well as horizontal and vertical angles.

GONIOMETER - Measures horizontal and vertical angles.

GRADIOMETER Also known as Gradiometer level, it measures slight inclines and level lines-of-sight.

HELIOGRAPH - Signalling device used in triangulation surveys.

HELIOSTAT - Also known as a heliotrope, it was used to make survey points visible at long distances, particularly in triangulation surveys.

HORIZON, ARTIFICIAL - Assists in establishing a level line of sight, or "horizon". HYPSOMETER - Used to estimate elevations in mountainous areas by measuring the boiling points of liquids. This name was also given to an instrument which determined the heights of trees.

INCLINOMETER - Measures slopes and/or vertical angles.

LEVEL - Measures vertical distances (elevations). There are many kinds, including Cooke's, Cushing's, Gravatt. Dumpy, hand or pocket, wye, architect's, builder's, combination, water, engineer's, etc.

LEVELLING ROD - A tool used in conjunction with a levelling instrument. LEVELLING STAVES - Used in measuring vertical distances.

MINER'S COMPASS - Determines magnetic direction; also locates ore. MINER'S PLUMMET - A "lighted" plumb bob, used in underground surveying.

MINING SURVEY LAMP - Used in underground surveying for vertical and horizontal alignment.

OCTANT - For measuring the angular relationship between two objects. PEDOMETER - Measures paces for estimating distances. PERAMBULATOR - A wheel for measuring horizontal distances.

PHOTO-THEODOLITE - Determines horizontal and vertical positions through the use of "controlled" photographs.

PLANE TABLE - A survey drafting board for map-making with an alidade. PLUMB BOB - For alignment; hundreds of varieties and sizes. PLUMMETS - Same as plumb bob.

QUADRANT - For measuring the angular relationship between two objects. RANGE POLES - For vertical alignment and extending straight lines. SEMICIRCUMFERENTER - Measures magnetic directions and horizontal angles.

SEXTANTS - Measures vertical angles; there are many kinds, including box, continuous arc, sounding, surveying, etc.

SIGNAL MIRRORS - For communicating over long distances; used in triangulation surveys. STADIA BOARDS - For measuring distances; also known as stadia rods.

STADIMETER or STADIOMETER - For measuring distances.

TACHEOMETER - A form of theodolite that measures horizontal and vertical angles, as well as distances.

TAPES - For measuring distances; made of many materials, including steel, invar, linen, etc. Also made in many styles, varieties, lengths, and increments.

THEODOLITE - Measures horizontal and vertical angles. Its name is one of the most misused in surveying instrument nomenclature, and is used on instruments that not only measure angles, but also directions and distances. There are many kinds, including transit, direction, optical, solar, astronomic, etc.

TRANSIT - For measuring straight lines. Like the theodolite, the transit's name is often misused in defining surveying instruments. Most transits were made to measure horizontal and vertical angles and magnetic and true directions. There are many kinds, including astronomic, solar, optical, vernier, compass, etc.

WAYWISER - A wheel for measuring distances.

2.3. Traverse (Surveying)

Traverse is a method in the field of surveying to establish control networks. It is also used in geodetic work. Traverse networks involved placing the survey stations along a line or path of travel, and then using the previously surveyed points as a base for observing the next point. Traverse networks have many advantages of other systems, including:

Less reconnaissance and organization needed.

While in other systems, which may require the survey to be performed along a rigid polygon shape, the traverse can change to any shape and thus can accommodate a great deal of different terrains.

Only a few observations need to be taken at each station, whereas in other survey networks a great deal of angular and linear observations need to be made and considered.

Traverse networks are free of the strength of figure considerations that happen n triangular systems.

Scale error does not add up as the traverse is performed. Azimuth swing errors can also be reduced by increasing the distance between stations.

The traverse is more accurate than triangulateration (a combined function of the triangulation and trilateration practice).

2.3.1. Types

Frequently in surveying engineering and geodetic science, control points (CP) are setting/observing distance and direction (bearings, angles, azimuths, and elevation). The CP throughout the control network may consist of monuments, benchmarks, vertical control, etc.

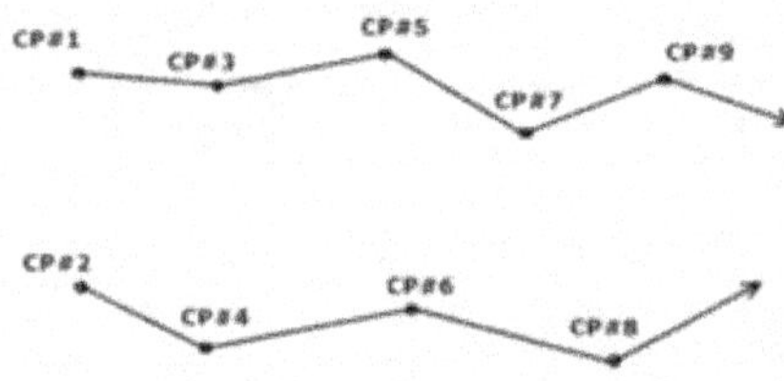

Figure 2.3.1: Diagram of an Open Traverse

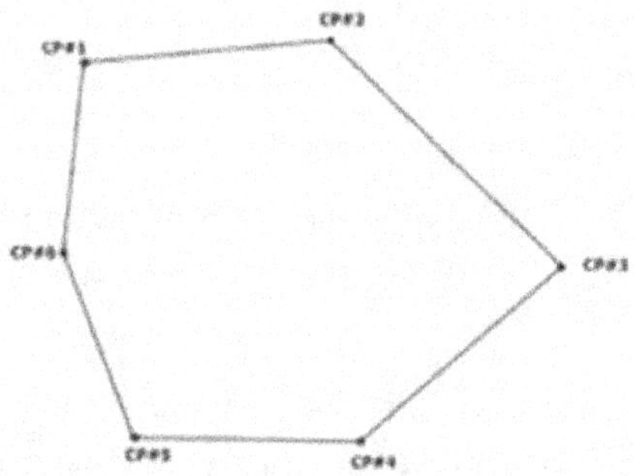

Figure 2.3.2: Diagram of a Closed Traverse

Open/Free

An open, or free traverse (link traverse) consist of known points plotted in any corresponding linear direction, but do not return to the starting point or close upon a point of equal or greater order accuracy. It allows geodetic triangulation for sub-closure of three known points; known as the "Bowditch rule" or "compass rule" in geodetic science and surveying, which is the principle that the linear error is proportional to the length of the side in relation to the perimeter of the traverse.

Closed

A closed traverse (polygonal, or loop traverse) is a practice of traversing when the terminal point closes at the starting point. The control points may envelop, or are set within the boundaries, of the control network. It allows geodetic triangulation for sub-closure of all known observed points.

Closed traverse is useful in marking the boundaries of wood or lakes. Construction and civil engineers utilize this practice for preliminary surveys of proposed projects in a particular designated area. The terminal (ending) point closes at the starting point.

2.3.2. Components and their Functions of Theodolite

A compass measures the direction by measuring the angle between the line and a reference direction, which is the magnetic meridian. A compass can measure angles up to an accuracy of 30 and by judgement up to an accuracy of 15. The principle of working of the compass is based on the property of the magnetic needle, which when freely suspended, takes the north-south direction. Compass measurements are thus affected by external magnetic influences and therefore a compass is unsuitable in some areas. In this here, we will discuss another method of measuring directions of lines; a theodolite is very commonly used to measure angles in survey work.

There are a variety of theodolites-Vernier, optic, electronic, etc. The improvements (from one form to the other) have been made to ensure ease of operation, better accuracy, and speed. Electronic theodolites display and store angles at the press of a button. This data can also be transferred to a computer for further processing. We start our discussion with the simplest theodolite-the Vernier theodolite.

The Vernier theodolite is a simple and inexpensive instrument but very valuable in terms of measuring angles. The common Vernier theodolite measures angles up to an accuracy of 20 in a compass, where the line of sight is simple, restricting its range, theodolites are provided with telescopes which provide for much greater range and better ac-curacy in sighting distant objects. It is, however, a delicate instrument and needs to be handled carefully. The theodolite measures the horizontal angles between lines and can also measure vertical angles. The horizontal angle measured can be the included angle, deflection angle or exterior angle in a traverse. The vertical angle is the angle in a vertical plane between the inclined line of sight of the instrument and the horizontal. In the following sections we will discuss the Vernier theodolite as well as its applications in surveying.

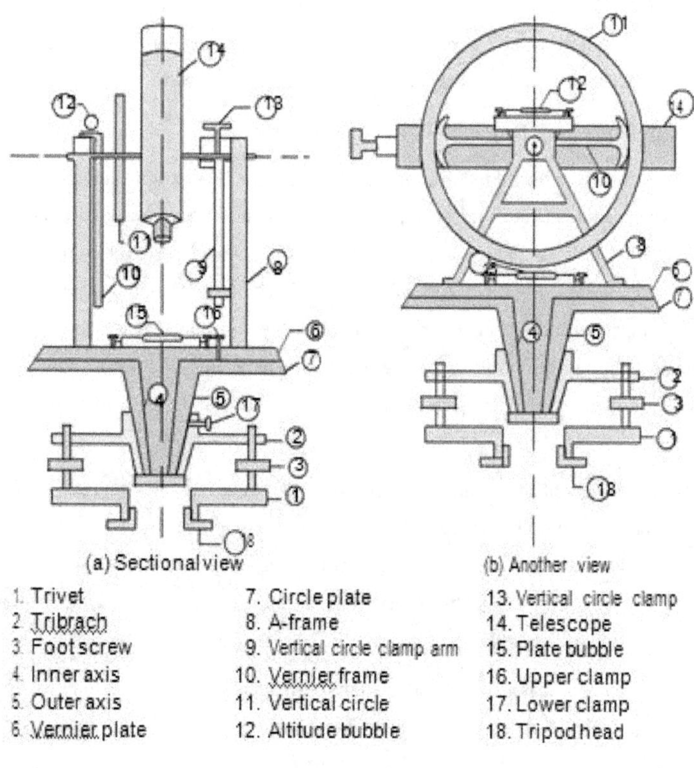

Figure 2.3.3: Vernier Theodolite

The vernier theodolite is also known as a transit. In a transit theodolite or simply transit the telescope can be rotated in a vertical plane. Earlier versions of theodolites were of the non-transit type and are obsolete now.

2.4. Vernier Theodolite

The vernier theodolite is also known as a *transit*. In a transit theodolite or simply transit the telescope can be rotated in a vertical plane. Earlier versions of theodolites were of the non-transit type and are obsolete now. Only the transit theodolite will be discussed here.

Two different views of a vernier theodolite are shown in Figs 2.4(a) and (b). The instrument details vary with different manufacturers but the essential parts remain the same. The main parts of a theodolite are the following.

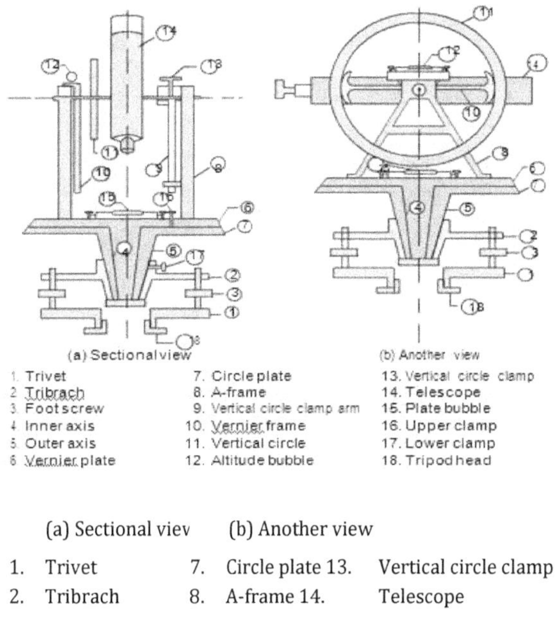

(a) Sectional view   (b) Another view

1. Trivet	7. Circle plate	13. Vertical circle clamp
2. Tribrach	8. A-frame	14. Telescope
3. Foot screw	9. Vertical circle clamp arm	15. Plate bubble
4. Inner axis	10. Vernier frame	16. Upper clamp
5. Outer axis	11. Vertical circle	17. Lower clamp
6. Vernier plate	12. Altitude bubble	18. Tripod head

Fig. 2.4: Vernier Theodolite

Levelling head The levelling head is the base of the instrument. It has the provision to attach the instrument to a tripod stand while in use and attach a plumb bob along the vertical axis of the instrument. The levelling head essentially consists of two triangular plates kept a distance apart by levelling screws. The upper plate of the levelling head, also known as the *tribrach*, has three arms, each with a foot screw. Instruments with four foot screws for levelling are also available. In terms of wear and tear, the three-foot-screw instrument is preferable. The lower plate, also known as the *trivet*, has a central hole and a hook to which a plumb bob can be attached. In modern instruments, the base plate of the levelling head has two plates which can move relative to each other. This allows a slight movement of the level- ling head relative to the tripod. This is called a *shifting head* and helps in centring the instrument over the station quickly. The functions of the levelling head are to support the upper part of the instrument, attach the theodolite to a tripod, attach a plumb bob, and help in levelling the instrument with the foot screws.

Lower plate the lower plate, also known as the *circle plate*, is an annular, Horizontal plate with a bevelled graduated edge fixed to the upper end of a hollow cylindrical part. The graduations are provided all around, from 0 o to 360°, in the clockwise direction. The graduations are in degrees divided into three parts so that each division equals 20″. An axis through the centre of the plate is known as the outer axis or the centre. Horizontal angles are measured with this plate. The diameter of the lower plate is sometimes used to indicate the size of or designate the instrument; for example, a 100-mm theodolite.

Upper plate The upper plate is also a horizontal plate of a smaller diameter attached to a solid, vertical spindle. The bevelled edge of the horizontal part carries two verniers on diametrically opposite parts of its circumference. These verniers are generally designated A and B. They are used to read fractions of the horizontal circle plate graduations. The centre of the plate or the spindle is known as the inner axis or centre. The upper and lower plates are enclosed in a metal cover to prevent dust accumulation. The cover plate has two glass windows longer than the vernier length for the purpose of reading. Attached to the cover plate is a metal arm hinged to the centre carrying two magnifying glasses at its ends. The magnifying glasses are used to read the graduations clearly.

Two axes or centres The inner axis as mentioned earlier is the axis of the conical spindle attached to the upper or vernier plate. The outer axis is the centre of the hollow cylindrical part attached to the lower or circle plate. These two axes coincide and form the vertical axis of the instrument, which is one of the fundamental lines of the theodolite.

Clamps and tangent screws There are two clamps and associated tangent or slow-motion screws with the plates. The clamp screws facilitate the motion of the instrument in a horizontal plane. The lower clamp screw locks or releases the lower plate. When this screw is unlocked, the lower and upper plates move together. The associated lower tangent screw allows small motion of the plates in the locked position. The upper clamp screw locks or releases the upper vernier plate. When this clamp is released (with the lower clamp locked), the lower plate does not move but the vernier plate moves with the instrument. This causes a change in the reading. The upper tangent screw allows for a small motion of the vernier plate for fine adjustments. When both the clamps are locked, the instrument cannot move in the horizontal plane. The construction of the clamp and tangent screws is shown in Fig. 2.4.1.

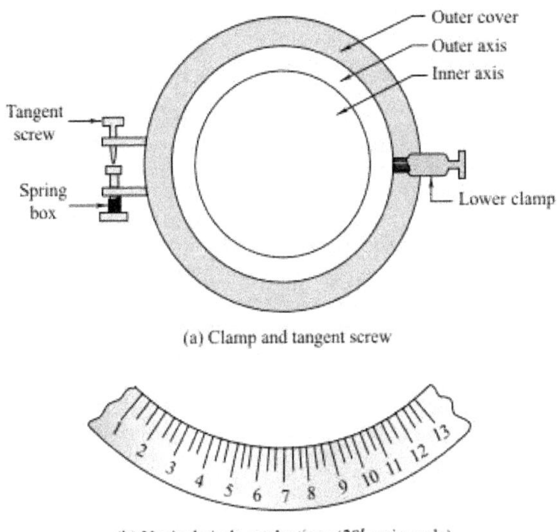

(a) Clamp and tangent screw

(b) Vertical circle graduations (20′ main scale)

Figure 2.4.1

Plate level The plate level is a spirit level with a bubble and graduations on the glass cover. A single level or two levels fixed in perpendicular directions may be provided. The spirit level can be adjusted with the foot screw of the levelling head. The bubble of the spirit level can be moved with the foot screws of the levelling head, which is a very fundamental adjustment required for using the theodolite. A small circular bubble may be provided for rough adjustment before levelling.

Index frame The index frame, also known as a T-frame or vernier frame, is a T-shaped metal frame. The horizontal arm carries at its ends two verniers, which remain fixed in front of the vertical circle. These verniers are generally designed C and D. The vertical leg of the T-frame, known as the clipping arm, has clipping screws with which the frame can be titled. The altitude level is generally fixed on top of this frame. When the telescope is rotated in a vertical plane, the vertical circle moves and vertical angles are measured on the vertical circle with the help of these verniers.

Standard or A-frame Two standards in the shape of the letter A are attached to the upper plate. The horizontal axis of the instrument is attached to these standards. The clipping arm of the index frame and the arm of the vertical circle clamp are also attached to the A-frame. The A-frame supports the telescope and the vertical circle.

Telescope The telescope is a vital part of the instrument. It enables one to see stations that are at great distances. The essential parts of a telescope are the eye-piece, diaphragm with cross hairs, object lens, and arrangements to focus the tele-scope. A focusing knob is provided on the side of the telescope. Earlier, external focusing telescopes were used. Today, only internal focusing telescopes are used in theodolites. These reduce the length of the telescope. The telescope may carry a spirit level on top in some instruments.

Vertical circle The vertical circle is a circular plate supported on the trunnion or horizontal axis of the instrument between the A-frames. The vertical circle has a bevelled edge on which graduations are marked. The graduations are generally quadrantal, 0° -90° in the four quadrants as shown in 2.4.1. The full circle system of graduations can also be seen in some instruments. The vertical circle moves with the telescope when it is rotated in a vertical plane. A metal cover is provided to protect the circle and the verniers from dust. Two magnifying glasses on metal arms are provided to read the circle and verniers. The cover has glass or plastic windows on which the magnifiers can be moved.

Vertical circle clamp and tangent screw The vertical circle is provided with a clamp and tangent screw as in the case of the horizontal plate, Upon clamping the vertical circle, the telescope cannot be moved in a vertical plane. The tangent screw allows for a slow, small motion of the vertical circle.

Altitude level is used for levelling, particularly when taking vertical angle observations. A circular or trough magnetic compass is generally fitted to the theodolite for measuring the magnetic bearing of lines. It is fitted on the cover of the horizontal plates. Two plates with graduations are provided in the compass box for ensuring that the needle ends are centred. The needle can be locked or released by a pin. When released, the telescope can be turned in azimuth to make the north end of the needle point to the north by making it read zero.

Tripod One accessory essential with the theodolite is the tripod on which it is mounted when it has to be used. The tripod head is screwed onto the base or the lower part of the levelling head. Its legs should be spread out for stability. The legs of the tripod are also used for rough levelling.

Plumb bob A heavy plumb bob on a good string with a hook at the end is required for centring the theodolite over a station. The plumb bob is fixed to the hook or other device projecting from the centre of the instrument in a central opening in the levelling head.

Main circle and vernier graduations In most of the instruments, the vernier enables readings up to 20" of the arc. This is made possible by marking the graduations on the circle and the vernier suitably as follows. The main circle is graduated into degrees and each degree is divided into three parts. Each main scale division thus represents 2". For the vernier, 59 main scale divisions are taken and divided into 60 parts. 59 main scale divisions form 59' 20". Therefore, each vernier scale division represents 59' 20/60 minutes. As you would have studied earlier, least count of the vernier = difference between a main scale division and a vernier scale division = main scale division - vernier scale division. Hence, in this case,

Least count = 20" – 59' 20/60 = 1/3 = (1/3) 60" = 20" Thus the least count of the vernier in common theodolites is 20".

Terminology of Theodolite

It is important to clearly understand the terms associated with the theodolite and its use and meaning. The following are some important terms and their definitions.

Vertical axis It is a line passing through the centre of the horizontal circle and perpendicular to it. The vertical axis is perpendicular to the line of sight and the trunnion axis or the horizontal axis. The instrument is rotated about this axis for sighting different points.

Horizontal axis It is the axis about which the telescope rotates when rotated in a vertical plane. This axis is perpendicular to the line of collimation and the vertical axis.

Telescope axis It is the line joining the optical centre of the object glass to the centre of the eyepiece.

Line of collimation It is the line joining the intersection of the cross hairs to the optical centre of the object glass and its continuation. This is also called the line of sight.

Axis of the bubble tube It is the line tangential to the longitudinal curve of the bubble tube at its centre.

Centring Centring the theodolite means setting up the theodolite exactly over the station mark. At this position the plumb bob attached to the base of the instrument lies exactly over the station mark.

Transiting It is the process of rotating the telescope about the horizontal axis through 180 o. The telescope points in the opposite direction after transiting. This process is also known as *plunging* or *reversing.*

Swinging It is the process of rotating the telescope about the vertical axis for the purpose of pointing the telescope in different directions. The right swing is a rotation in the clockwise direction and the left swing is a rotation in the counter-clockwise direction.

Face-left or normal position This is the position in which as the sighting is done, the vertical circle is to the left of the observer.

Face-right or inverted position This is the position in which as the sighting is done, the vertical circle is to the right of the observer.

Changing face It is the operation of changing from face left to face right and vice versa. This is done by transiting the telescope and swinging it through 180°.

Face-left observation It is the reading taken when the instrument is in the normal or face-left position.

Face-right observation It is the reading taken when the instrument is in the inverted or face- right position.

2.5. Method of Repetition in Theodolite

In the method of repetition, the horizontal angle is measured a number of times and the average value is taken. It is usual to limit the number of repetitions to three with each face except in the case of very precise work. With large number of repetitions, errors can also increase due to bisections, reading the verniers, etc. Very large number of repetitions necessarily do not lead to a more precise value of the angle. However, a number of errors are eliminated by the repetition method. The procedure is as follows (Fig. 2.5.1).

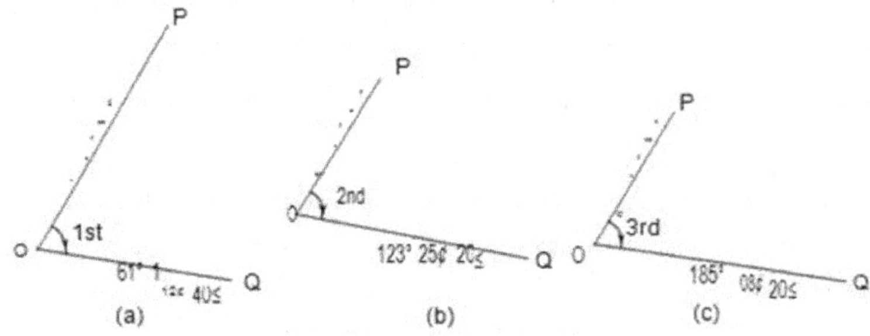

Figure 2.5.1: Repetition Method

1. Angle POQ is to be measured. Set up, centre, and level the theodolite at O. Ensure that the instrument is in the normal position, i.e., face left.

2. Set the instrument to read 0°00′ 00″. For this release the upper clamp and bring the zero of the vernier (at vernier A) very close to the zero of the circle. Clamp the upper plate and using the upper tangent screw, coincide the two zeros exactly.

3. Loosen the lower clamp and rotate the instrument so that the left signal at P is approximately bisected. Tighten the lower clamp and using the lower tangent screw, bisect the signal at P exactly. Read the verniers at A and B. The reading should not change and they should read zero and 180°.

4. Loosen the upper clamp and rotate the instrument clockwise to bisect the right signal at Q. Using the upper tangent screw, bisect the signal at approximately Q exactly.

5. Read the verniers at A and B. The reading at A gives the value of the angle directly. The reading on the vernier at B will be 180° + the angle. Record both the readings.

Release the lower clamp and rotate the instrument clockwise to bisect the signal at the left station P again. Using the lower tangent screw, bisect the signal.

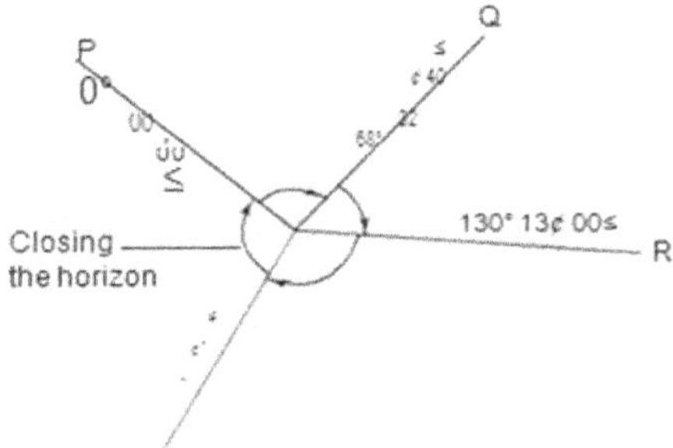

Figure 2.5.2

The method of repetition helps to eliminate the following errors.

1. Errors caused by the eccentricity of the centres and verniers, by reading both the verniers and averaging.

2. Graduation errors by reading from different parts of the circle.

3. Imperfect adjustment of the line of collimation and horizontal axis by face-left and face- right observations.

4. Observational errors and other errors tend to be compensated by the large number of readings.

However, the errors due to levelling cannot be compensated. This has to be done by permanent adjustment. Also a large number of repetitions tend to increase the wear of clamp and tangent screws.

Therefore, from the two sets,

Mean value of the angle = $(1/2)(61°42'47" + 61°42'40") = 61°42'44"$

2.5.1. Locating Landscape Details with the Theodolite

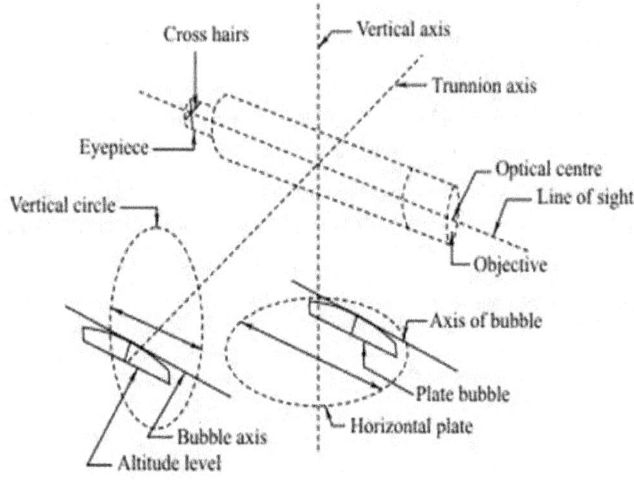

Figure 2.5.3: Fundamental Lines of a Theodolite

In most surveys, it is necessary to locate details such as buildings, railway lines, canals, and other landmarks along with the survey. A transit with a steel tape is used to locate details, and many methods are available, as the transit is an angle-measuring instrument.

Locating Landscape Details with the Theodolite

We have discussed so far methods to survey the main frame or the skeleton of the survey. In most surveys, it is necessary to locate details such as buildings, railway lines, canals, and other landmarks along with the survey. A transit with a steel tape is used to locate details, and many methods are available, as the transit is an angle-measuring instrument. The following methods can be used.

Angle and Distance from a Single Station

A point can be located with an angle to the station along with the distance from that station. The angle is preferably measured from the same reference line to avoid confusion. A sketch with the line and the distance and angle measured will help in plotting later. Angles to a number of points are measured and with each angle two distances are measured to locate the road.

Angle from One Station and Distance from Another

If for any reason, it is not possible to measure the angle and distance to an object from the same point, it may be possible to locate the point by measuring angles from one station and distances from the other. The recorded data should clearly indicate the stations from which the angle and distance are measured. The angle is measured from station A to point P. When the instrument is shifted to B, the distance to point P is measured from B with a steel tape.

Angles from two Stations

If for some reason, it is not possible to measure distances, then angles from two stations are enough to locate a point. The point P is located by measuring angles to point P from stations A and B.

The following are the fundamental lines.

1. The vertical axis
2. The horizontal or trunnion axis
3. The line of collimation or line of sight
4. Axis of altitude level
5. Axis of plate level

The meaning of these terms has been discussed earlier. The axes are shown in Fig. 2.5.4. When the instrument is properly adjusted, the relationships between these axes are the following.

a. The horizontal axis must be perpendicular to the vertical axis.
b. The axis of the plate level must be perpendicular to the vertical axis.
c. The line of collimation must be at right angles to the horizontal axis.
d. The axis of the altitude level (and telescope level) must be parallel to the line of collimation.
e. The vertical circle vernier must read zero when the line of sight is horizontal. Each one of these relations gives conditions for accurate measurement.
   1. When the horizontal axis is perpendicular to the vertical axis, the line of sight generates a vertical plane when transited.
   2. When the axis of the plate level is perpendicular to the vertical axis, the vertical axis will be truly vertical when the bubble traverses.

3. When the line of collimation is at right angles to the horizontal axis, the tele-scope when rotated about the horizontal axis will move in a vertical plane.

4. When the line of collimation and the axis of altitude level are parallel, the vertical angles will be measured without any index error.

5. The index error due to the displacement of the vernier is eliminated when the vernier reads zero with the line of collimation truly horizontal.

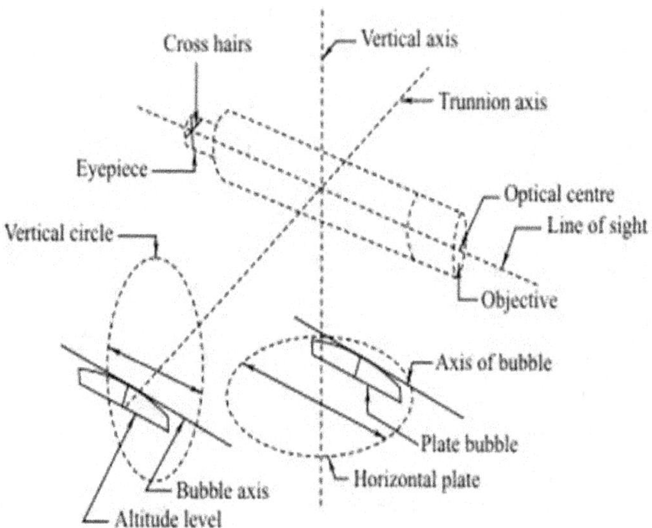

Figure 2.5.4: Fundamental Lines of a Theodolite

2.6. Tacheometric Systems

- Tacheometry (or tachometry or telemetry) is a branch of angular surveying in which the horizontal and vertical distances of points are obtained by optical means as opposed to the ordinary slower process of measurements by tape or chain.

- The method is very rapid and convenient.

- The accuracy of Tacheometry in general compares unfavourably with that of chaining, it is best adapted in obstacles such as steep and broken ground, deep ravines, stretches of water or swamp and so on, which make chaining difficult or impossible.

- The accuracy attained is such that under favourable Conditions the error will not exceed 1/1000.

- The primary object of Tacheometry is the preparation of contoured maps or plans requiring both the horizontal as well as vertical control.
- Also, on surveys of 'higher accuracy, it provides a check on distances measured with the tape.

An ordinary transit theodolite fitted with a stadia diaphragm is generally used for tacheometric survey. the stadia diaphragm essentially consists of one 'stadia hair above and the other an equal distance below the horizontal cross-hair, the stadia hairs being mounted in the same ring and in the same vertical plane as the horizontal and vertical cross-hairs. Fig. 2.6.1 shows the different forms of stadia diaphragm commonly used.

The telescope used in stadia surveying are of three kinds.

(1) The simple external-focusing telescope.

(2) The external-focusing anallactic telescope (porro's telescope).

(3) The internal-focusing telescope.

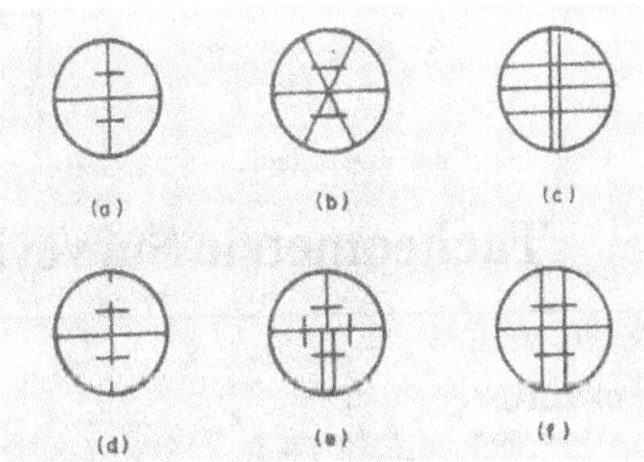

Figure 2.6.1

The first type is known as stadia theodolite, while the second is known as tacheometer the 'tacheometer' (as such) has the advantage over the first and the third type due to the fact that the additive constant of the instrument is zero.

However, the internal focusing telescope is becoming more popular, though it has a very small additive constant.

A tacheometer must essentially incorporate the following features:

(i) The multiplying constant should have a nominal value of 100 and the error contained in this value should not exceed 1 in 1000.

(ii) The axial horizontal line should be exactly midway between the other two lines.

(iii) The telescope should be truly anallactic.

(iv) The telescope should be powerful having a magnification of 20 to 30 diameters.

The aperture of the objective should be 35 to 45 mm in diameter in order to have a sufficiently bright image.

For small distances (say up to 100 meters), ordinary leveling staff may be used. For greater distances a stadia rod may be used. A stadia rod is usually of one piece, having 3 to 5 meters length.

The pattern of graduations should be bold and simple. For stadia work done, the finer graduations are usually omitted. For smaller distances, a stadia rod graduated in 5 mm (le. 0.005 m) may be used, while for longer distances, the rod may be graduated in 1 cm (le. 0.01 M).

2.6.1. *Different Systems of Tacheomefric Measurement*

The various systems of tachometric survey may be classified as follows:

(1) The stadia system

    (a) Fixed hair method

    (b) Movable hair method or subtense method.

(2) The tangential system.

(3) Measurements by means of special instruments.

2.7. Principle of Stadia Method

The stadia method is based on the principle that the ratio of the perpendicular to the base is constant in similar isosceles triangles.

In Fig.2.7.1, let two rays OA and OR be equally inclined to the central ray OC. Let A B A B and AS be the staff intercepts. Evidently $OC_2/A_2B_2 = OC_1/A_1B_1 = OC/AB$

$$K = 1/2 \cot ß/2$$

This constant k entirely depends upon the magnitude of the angle ß. If ß is made equal to 34′ 22″, the constant k= cot 17′ 11″ = 100. In this case, the distance between the staff and the point 0 will be 100 times the staff intercept.

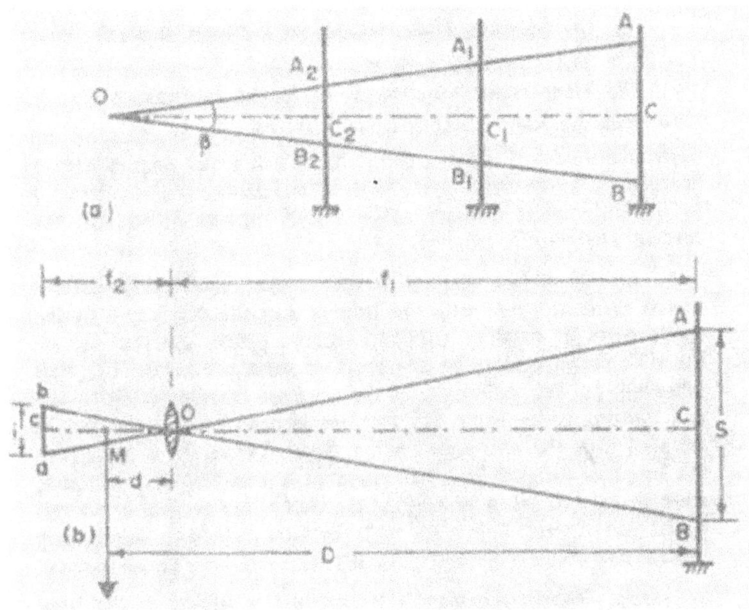

Figure 2.7.1: Principle of Stadia Method

In Fig. 2.7.1, let two rays OA and OR be equally inclined to the central ray OC. Let A B A B and AS be the staff intercepts. Evidently $OC_2/A_2B_2 = OC_1/A_1B_1 = OC/AB$

$$K = 1/2 \cot ß/2$$

This constant k entirely depends upon the magnitude of the angle ß. If ß is made equal to 34' 22", the constant k= cot 17' 11" = 100. In this case, the distance between the staff and the point 0 will be 100 times the staff intercept.

Distance from the axis to the staff is given by,

$$D = FC + (f + d)$$
$$= f/I \, s + (f+d)$$
$$= k \, s + C$$

2.7.1. Distance and Elevation Formulae for Staff Vertical

Inclined Sight

Let

P = Instrument station

Q = Staff station

M = Position of instruments axis

O = Optical centre of the objective

A, C, B = Points corresponding to the readings of the three hairs

s = AB= Staff intercept

i = Stadia interval

θ = Inclination of the line of sight from the horizontal

L = Length MC measured along the line of sight

D '= MQ' = Horizontal distance between the instrument and the staff

V = Vertical intercept, at Q, between the line of sight and the horizontal Line.

h = Height of the instrument

r = Central hair reading

f = Angle between the two extreme rays corresponding to stadia hairs.

Draw a line A'CB' (Fig. 2.7.2) normal to the line of sight OC.

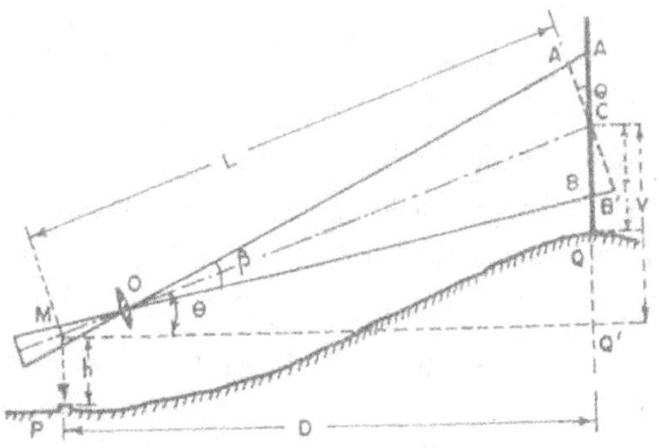

Figure 2.7.2: Elevated Sight: Vertical Holding

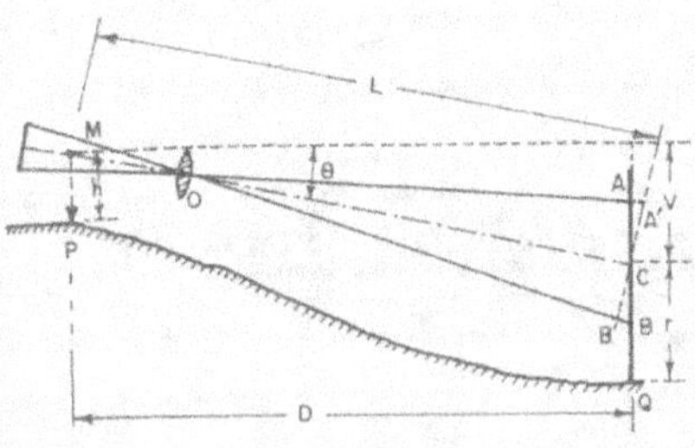

Figure 2.7.3: Depressed Sight: Vertical Holding

Distance and Elevation Formulae for Staff Normal

Case (a) Line of sight at an angle of elevation θ

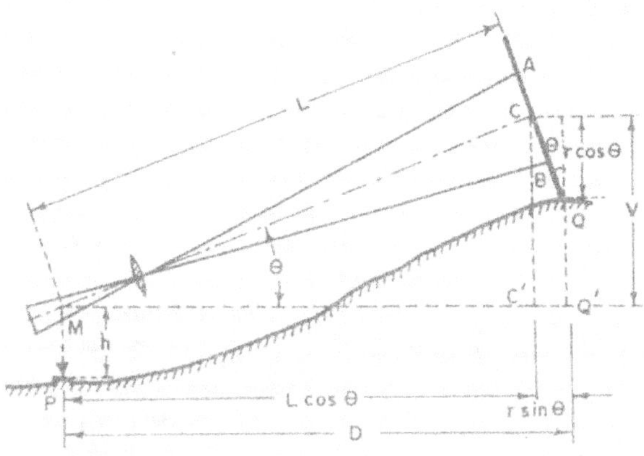

Figure 2.7.4: Elevated Sight: Normal Holding

Let AB = s = Staff intercept

CQ = r = Axial hair reading

With the same notations as in the previous case, we have

MC = L = ks + C

The horizontal distance between P and Q is given by

D =MC′ + C′Q′ =L cos θ +r sin θ

= (k + C) cos θ + r sin θ

Similarly, V = L sin θ = (k s + C) sin θ

Elev. of Q=Elev. of P+ h + V—r cos θ

Case (b) Line of sight at an angle of depression θ

When the line of sight is depressed downwards,

$$MC = L = ks + C$$
$$D = MQ' = MC' - Q'C' = L \cos \theta - r \sin \theta$$
$$= (ks + C) \cos \theta - r \sin \theta \qquad \qquad ...(1.8)$$
$$V = L \sin \theta = (ks + C) \sin \theta \qquad \qquad ...(1.9)$$
$$\text{Elev. of } Q = \text{Elev. of } P + h - V - r \cos \theta$$

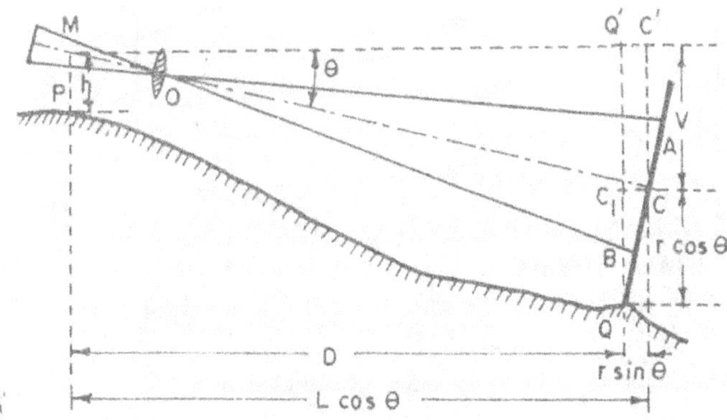

2.7.2. Anallactic Telescope Inclined Sight

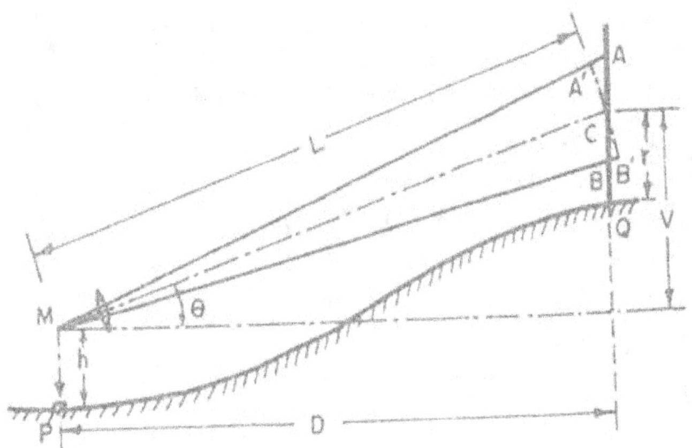

Figure 2.7.5: Anallactic Lens: Inclined Sight

With the same notations as that of figure 2.7.5, we have

$$MC = L = k \cdot A'B' = ks \cos \theta \qquad \text{...(i)}$$

$$D = L \cos \theta = ks \cos^2 \theta \qquad \text{...(ii)}$$

and
$$V = L \sin \theta = ks \cos \theta \sin \theta = \frac{ks}{2} \sin 2\theta \quad \text{...(ii)}$$

R.L. of
$$Q = \text{R.L. of } P + h + V - r$$

2.8. Stadia Constants

The instrumental constants k and c can best be determined by measuring the additive constant c along the telescope (as in the case of stadia method) and observing the micrometer readings corresponding to staff kept at some measured distance.

Let D1 and D2 = Measured distances from the instrument s = Fixed distance between the two targets.

m1= Sum of the two micrometer readings when the staff is kept at distance D. m2= Sum of two micrometer readings when the staff is kept at distance D2.

e = Index error.

Substituting the corresponding values in equation 1.13 (a), we get.

$$D_1 = \frac{K.s}{m_1 - e} + C$$

or
$$K.s = (D_1 - C)(m_1 - e)$$

and
$$D_2 = \frac{K.s}{m_2 - e} + C$$

or
$$K.s = (D_2 - C)(m_2 - e)$$

Equating (i) and (ii), we get

$$(D_1 - C)(m_1 - e) = (D_2 - C)(m_2 - e)$$

From which
$$e = \frac{(D_2 - C)m_2 - (D_1 - C)m_1}{(D_2 - D_1)}$$

Substituting the value of e in (i), we get

$$K = \frac{D_1 - C}{s}\left\{ m_1 - \frac{(D_2 - C)m_2 - (D_1 - C)m_1}{(D_2 - D_1)}\right\}$$

or
$$K = \frac{D_1 - C}{s(D_2 - D_1)}\left(m_1 D_2 - m_1 D_1 - m_2 D_2 + m_2 C + m_1 D_1\right.$$

or
$$K = \frac{(D_1 - C)(D_2 - C)(m_1 - m_2)}{s(D_2 - D_1)}$$

Example 1: A reservoir of bottom size 35mx 15m planned with a depth of 4 m. The slide slope is 1.5:1. Calculate the quantity to be excavated assuming the surface of the ground to be leveled before excavation.

Given Data

The depth of cutting (h) = 4 m

Slide slope = 1.5:1

S = 1.5

Solution

Length of excavation required top = L+2sh

= 35 +2 (1.5)

= 47 m

Width of excavation required top = b +2sh

= 25+2(1.5x4)

=37 m

To find average length and breath

Average length = 47+35/2

= 41 m

Average width= 37+25/2

= 31 m

Area at top (At) = 47x37 = 1739 m²

Area at centre (Ac) = 41x31 = 1271 m²

Area at bottom (Ab) = 25x35 = 875 m²

Using prismodial rules the volume of excavation = 2/3 [dx (At+Ac+Ab)]

=2/3[4(875+1739+1271)]

= 513 m³

MODEL SHORT QUESTION AND ANSWERS

THEODOLITE AND TACHEOMETRIC SURVEYING

1)   **What are the uses of a contour map? (April/May 2018) (April/May 2011) (April/May 2017)**

Contour maps are very useful since they provide valuable information about the terrain. Some of the uses are as follows:

- The nature of the ground and its slope can be estimated. Earth work can be estimated for civil engineering projects like road works, railway, canals, dams etc.

It is possible to identify suitable site for any project from the contour map of the region. Inter-visibility of points can be ascertained using contour maps. This is most useful for locating communication towers.

2) What are the various methods of doing the theodolite traversing? (April/May 2018)

1. Magnetic bearing method
2. Loose needle method
3. Direct angle method
4. Included angle method
5. Deflection angle method

3) What are the various methods of doing the tachometric surveying? (April/May 2018) (April/May 2016)

- There are two methods of tachometry survey. They are Stadia system, and tangential system.

4) State any two characteristics of contour. (Nov/Dec 2017)

- All points on a contour line are of the same elevation.
- No two contour lines can meet or cross each other except in the rare case of an overhanging vertical cliff or wall. Closely spaced contour lines indicate steep slope.
- Widely spaced contour lines indicate gentle slope.
- Equally spaced contour lines indicate uniform slope.
- Closed contour lines with higher elevation towards the centre indicate hills.
- Closed contour lines with reducing levels towards the centre indicate pond or other depression.
- Contour lines of ridge show higher elevation within the loop of the contours. Contour lines cross ridge at right angles.

5) Define Standards of theodolite. (Nov/Dec 2017)

- Theodolite has many parts which needs to be adjusted every time while surveying. It is important to know about theodolite parts and their functions before using it to minimize errors during theodolite surveying.
- Theodolite is an instrument used in surveying to measure horizontal and vertical angles. It is also used for leveling, indirect measure of distances and prolonging a line etc. The line of sight of theodolite can be rotated through 180° in vertical plane about its horizontal axis.

6) **What are the uses of a tacheomertry? (April/May 2017)**

- The primary objective of this tacheometric surveying is to prepare contoured maps or plans requiring both the horizontal as well as vertical control.
- On surveys of higher accuracy, it provides a check on distances measured with the tape.
- Preparation of topographic maps which require both elevations and horizontal distances.
- Survey work in difficult terrain where direct methods are inconvenient.
- Reconnaissance surveys for highways, railways, etc.
- Checking of already measured distances.
- Hydro graphic surveys.

7) **Define Contour intervals. (Nov/Dec 2017) (Nov/Dec 2014)**

- Contour interval means the difference in altitude represented by the space between two contour lines on a map.

8) **What are the different types of telescope used in stadia surveying? (Nov/Dec 2016)**

Based on various types of instruments used, surveying can be classified into six types.

1. Chain surveying
2. Compass surveying
3. Plane table surveying
4. Theodolite surveying
5. Tacheometric surveying
6. Photographic surveying

9) **What are the various methods of locating countour? (Nov/Dec 2016)**

There are two methods of contour surveying:

- Direct method
- Indirect method

10) **What are the three types of telescopes used in stadia surveying? (Nov/Dec 2016) (Nov/Dec2009)**

- External focusing telescope.
- External focusing analytic telescope.
- Internal focusing telescope.

11) **What is meant face left and face right? (April/May 2016)**

- When the vertical circle of the theodolite is on the left of the observer at the time of observation is called Face left.

- When the vertical circle of the theodolite is on the right of the observer at the time of observation is called Face right.

12)   What are the two methods of measuring the horizontal angle using a Theodolite? When each method is advantageously used?

Repetition Method

Reiteration Method

The method of repetition is preferred for the measurement of a single angle.

The method of reiteration is preferred in triangulation, where a number of angles may be required at one point by the instrument.

13)   What are the errors eliminated in measurements of horizontal angle by method of repetition?

Instrumental and Observational errors are eliminated in measurements of horizontal angle by method of repetition.

14)   What is transit Theodolite?

Transit theodolite is defined as the theodolite, in which its telescope can be rotated horizontally through 180° in the vertical plane.

15)   List out the major parts of a Theodolite.

1. Top Assembly (Alidade Assembly)
2. Middle Assembly (Horizontal Circle Assembly)
3. Bottom Assembly (Levelling Head Assembly)

16)   List the qualities of a Theodolite telescope?

Internal focusing telescopes are best suited instead of external telescopes. The magnification factor of the internal focusing telescope should be from 15 to 30 times of the diameter.

17)   What you mean by temporary adjustments of a Theodolite? (Nov/Dec 2014)

- The adjustments required to be made at every instrument station before taking observations are called temporary adjustments.
- The temporary adjustments of a theodolite consist of the following operations.

    1. Setting and centring the theodolite
    2. Levelling of the theodolite
    3. Elimination of parallax

18) What are the different systems of tacheometric surveying?

1) Stadia System

In the stadia system the diaphragm of the tacheometer is provided with two stadia hairs (upper & lower).

They are two kinds of stadia system.

Fixed hair method & movable hair method.

19) Tangential system

In the tangential system the diaphragm of the tacheometer is not provided with stadia hairs. Only the single horizontal hair is used to take the reading.

20) What is a collimation adjustment? Adjustment of the level of the Telescope

In this adjustment, the line of collimation should remain horizontal, when the bubble of the level tube fitted on telescope is brought at the centre of its run. This adjustment is essential when a theodolite is used as a level and also when vertical angles are observed.

21) What is mean by tacheometric surveying?

Tacheometric surveying is the branch of surveying in which angular observations are made with an instrument called tacheometer to determine horizontal and vertical distance. Here, use of chain has been completely eliminated.

22) Explain fixed hair method?

In this type of method the distance between the stadia hairs is fixed and thus the method is known as fixed hair method.

23) Explain movable hair method?

In this type of method the distance between the stadia hairs is varied while the staff intercept is kept constant. The staff is provided with two tangents at a known distance apart and a third target at a middle.

24) What is an analytic lens?

An analytic lens is an additional convex lens is mounted in external focusing telescope in between the object glass and diaphragm. This arrangement is made to reduce the additive constant zero. This arrangement simplifies the mathematical calculations and only multiplying constant is zero.

25)   What are the three types of telescopes used in stadia surveying?

    1)   External focussing telescope

    2)   External focussing anallactic telescope

    3)   internal focussing telescope

26)   Define height and distance?

Trigonometrical levelling is an indirect method of levelling. The relative elevation of various points are determined from the observed vertical angles and horizontal distance by the use of trigonometrical relations. The vertical angle are measured with the theodolite and horizontal are measured with the tape or chain.

27)   What are the different methods used to find the elevations of the points in the case of inaccessible points? Differentiate that?

Single plane method & Double plane method.

S. No	Single plane method	Double plane method
1	Two instrument stations are chosen in line with the object.	Two instrument stations are chosen which are not in line with the object.
2	Two vertical angles are measured in the same vertical plane	Two vertical angles are measured in the two vertical plane
3	Horizontal angles are not required	Horizontal angles are also measured

MODEL DETAIL QUESTIONS

1.   A theodolite was set up at a distance of 200m from a chimney and the angle of elevation to its top was $10^{0}48'$. The staff reading on BM of R.L 70.25m with the telescope horizontal was 0.977. Find the reduced level of the top of the Chimney. **(April/May 2018) (Nov/Dec 2017).**

2.   Two observation are taken upon a vertical staff by means of a theodolite of which the R.L of the horizondal axis is 254.30m. In case of the first, the line of sight is direct to give a staff reading of 1.00 and the angle of elevation is $4^{0}58'$. In the second observation the staff reading is 3.66m and the angle of elevations is $5^{0}44'$. Compute the R.L of staff station and its horizondal distance from the instrument. **(April/May 2018)**

3.   A theodolite was setup at a distance of 180m from a light house and the angle of elevation to its top and depression to its base were observed as $22^{0}45'$ and $1^{0}12'$ respectively. The reading on a staff held on B.M of R.L 175.590m was 1.85m with line of collimation horizontal. Calculate.

    1.   The height of lighthouse

    2.   The R.L of top **(April/May 2016) (April/May 2017).**

4. The following observations was taken to find the gradient of RS with a tacheometer provided with anallactic lens. **(Nov/Dec 2015) (April/May 2017).**

Inst. Station	Bearing	Top hair	Middle hair	Bottom hair	Vertical angle
R	340^0	0.900	1.722	2.544	$+15^0$
S	70^0	0.750	2.205	3.660	$+10^0$

5. Explain the method of reiteration for horizontal angle measurement. **(Nov/Dec 2013) (April/May 2017).**

6. What are the possible sources of error while using a theodolite? How can they be eliminated? **(April/May 2015).**

7. Two sets of tacheometer readings were taken from an instrument station A, the reduced levels of which was 100.06m to a staff station B. Instrument P-multiplying constant 100, additive constant 0.06m, staff held vertical. Instrument Q- multiplying constant 90, additive constant 0.06, staff held normal.

Instrument at to H.I vertical angle Stadia readings (m)

P   A B 1.50m 26^0 0.755, 1.005, 1.255
Q   A B 1.45m 26^0 ?

8. What should be the stadia readings with instrument Q?. **(Nov/Dec 2016).**

   (a) Explain angle measuring procedures using theodolite.

   (b) Explain the permanent adjustment of a theodolite. **(Nov/Dec 2016).**

9. What do you mean by contouring? Describe its characteristics with neat sketch and its uses. **(April/May 2017) (Nov/Dec 2016) (April/May 2015) (Nov/Dec 2014).**

10. Write a case study on contour mapping of hilly terrain. **(Nov/Dec 2016).**

UNIT III

3. CONTROL SURVEYING AND ADJUSTMENT

Horizontal and vertical control – Methods – specifications – triangulation- baseline – satellite stations – reduction to centre – trigonometrical levelling – single and reciprocal observations – traversing – Gale's table – Errors Sources – precautions and corrections – classification of errors – true and most probable values – weighed observations – method of equal shifts – principle of least squares – normal equation – correlates – level nets – adjustment of simple triangulation networks.

3. Control Surveying

3.1. Horizontal Controls & Its Methods

The horizontal control consists of reference marks of known plan position, from which salient points of designed structures may be set out. For large structures primary and secondary control points are used. The primary control points are triangulation stations. The secondary control points are reference to the primary control stations.

Reference Grid

Reference grids are used for accurate setting out of works of large magnitude. The following types of reference grids are used:

1. Survey Grid
2. Site Grid
3. Structural Grid
4. Secondary Grid

Survey grid is one which is drawn on a survey plan, from the original traverse. Original traverse stations form the control points of the grid. The site grid used by the designer is the one with the help of which actual setting out is done. As far as possible the site grid should be actually the survey grid. All the design points are related in terms of site grid coordinates. The structural grid is used when the structural components of the building are large in numbers and are so positioned that these components cannot be set out from the site grid with sufficient accuracy. The structural grid is set out from the site grid points. The secondary grid is established inside the structure, to establish internal details of the building, which are otherwise not visible directly from the structural grid.

3.2. Vertical Control & Its Methods

The vertical control consists of establishment of reference marks of known height relative to some special datum. All levels at the site are normally reduced to the nearby bench mark, usually known as master bench mark.

The setting of points in the vertical direction is usually done with the help of following rods:

1. Boning rods and travellers
2. Sight Rails
3. Slope rails or batter boards
4. Profile boards

A boning rod consist of an upright pole having a horizontal board at its top, forming a 'T' shaped rod. Boning rods are made in set of three, and many consist of three 'T' shaped rods, each of equal size and shape, or two rods identical to each other and a third one consisting of longer rod with a detachable or movable 'T' piece. The third one is called traveling rod or traveler.

Sight Rails

A sight rail consist of horizontal cross piece nailed to a single upright or pair of uprights driven into the ground. The upper edge of the cross piece is set to a convenient height above the required plane of the structure, and should be above the ground to enable a man to conveniently align his eyes with the upper edge. A stepped sight rail or double sight rail is used in highly undulating or falling ground. Slope rails or Batter boards:

These are used for controlling the side slopes in embankment and in cuttings. These consist of two vertical poles with a sloping board nailed near their top. The slope rails define a plane parallel to the proposed slope of the embankment, but at suitable vertical distance above it. Travelers are used to control the slope during filling operation.

Profile Boards

These are similar to sight rails, but are used to define the corners, or sides of a building. A profile board is erected near each corner peg. Each unit of profile board consists of two verticals, one horizontal board and two cross boards. Nails or saw cuts are placed at the top of the profile boards to define the width of foundation and the line of the outside of the wall.

- An instrument was set up at P and the angle of elevation to a vane4 m above the foot of the staff held at Q was 9º30?. The horizontal distance between P and Q was known to be2000 metres. Determine the R.L. of the staff station Q given that the R.L. of the instrument axis was 2650.38.

Solution

Height of vane above the instrument axis

=D tan?=2000 tan 9°30 '

=334.68 m

Correction for curvature and refraction

C =0.06735 Dinm, when D is in km

=0.2694? 0.27 m (+ve) Height of vane above the instrument axis

=334.68 +0.27 =334.95

R.L. fo vane=334.95+2650.38 =2985.33 m

R.L. of Q=2985.33- 4=2981.33 m

- An instrument was set up at P and the angle of depression to a vane 2 m above the foot of the staff held at Q was 5°36'. The horizontal distance between P and Q was known to be 3000 metres. Determine the R.L. of the staff station Q given that staff reading on a B.M. of elevation 436.050 was 2.865 metres.

Solution

The difference in elevation between the vane and the instrument axis.

= D tanθ

=3000 tan 5°36'=294.153

Combined correction due to cuvature and refraction

C =0.06735 D in metres, when D is in km

=0.606 m.

Since the observed angle is negative, the combined correction due to curvature and refraction is subtractive.

Difference in elevation between the vane and the instrument axis.

=294.153 -0.606 =293.547 =h. R.L. of instrument axis =436.050 +2.865 =438.915

R.L. of the vane=R.L. of instrument aixs-h =438.915 -293.547 =145.368

R.L. of Q=145.368-2 =143.368 m.

- In order to ascertain the elevation of the top (Q) of the signal on a hill, observations were made from two instrument stations P and R at a horizontal distance 100 metres apart, the station P and R being in the line with Q. The angles of elevation of Q at P and R were 28°42′ and 18°6′ respectively. The staff reading upon the bench mark of the elevation 287.28 was 2.870 and 3.750 respectively when the instrument was at P and at R, the telescope being horizontal. Determine the elevation of the foot of the signal if the height of the signal above its base is 3 metres.

Solution

Elevation of instrument axis at P=R.L. of B.M. + Staff reading

=287.28 +2.870 =290.15 m

Elevation of instrument axis at R=R.L. of B.M. +staff reading

=287.28 +3.750 =291.03 m

Difference in level of the instrument axes at the two stations

S =291.03 -290.15 =0.88 m.

=28°42′=18°6′

s cotθ=0.88 cot 18°6′=2.69 m

=152.1 m.

h--=D tanθ =152.1 tan 28°42′= 83.272 m

R.L. of foot of signal = R.L. of inst. aixs at P + h---ht. of signal

=290.15 +83.272 -3=370.422 m. Check: (b +D) =100 +152.1 m = 252.1 m

h--=(b + D) tanθ=252.1 x tan 18°6′

=82.399 m

R.L. of foot of signal =R.L. of inst. axis at R + h+ ht. of signal = 291.03 +82.399-3 = 370.429 m.

3.3. Triangulation

The basis of the classification of triangulation figures is the accuracy with which the length and azimuth of a line of the triangulation are determined. Triangulation systems of different accuracies depend on the extent and the purpose of the survey. The accepted grades of triangulation are:

1.   First order or Primary Triangulation.
2.   Second order or Secondary Triangulation.
3.   Third order or Tertiary Triangulation.

1) First Order or Primary Triangulation

The first order triangulation is of the highest order and is employed either to determine the earth's figure or to furnish the most precise control points to which secondary triangulation may be connected. The primary triangulation system embraces the vast area (usually the whole of the country). Every precaution is taken in making linear and angular measurements and in performing the reductions. The following are the general specifications of the primary triangulation:

1. Average triangle closure: Less than 1 second.

2. Maximum triangle closure: Not more than 3 seconds.

3. Length of base line: 5 to 15 kilometers.

4. Length of the sides of triangles: 30 to 150 kilometers.

5. Actual error of base: 1 in 300,000.

6. Probable error of base: 1 in 1,000,000.

7. Discrepancy between two measures of a section: 10 mm kilometres.

8. Probable error or computed distance: 1 in 60,000 to 1 in 250,000.

9. Probable error in astronomic azimuth: 0.5 seconds.

2) Secondary Order or Secondary Triangulation

The secondary triangulation consists of a number of points fixed within the framework of primary triangulation. The stations are fixed at close intervals so that the sizes of the triangles formed are smaller than the primary triangulation. The instruments and methods used are not of the same utmost refinement. The general specifications of the secondary triangulation are:

1. Average triangle closure: 3 sec

2. Maximum triangle closure: 8 sec

3. Length of base line: 1.5 to 5 km

4. Length of sides of triangles :8 to 65 km

5. Actual error of base: 1 in 150,000

6. Probable error of base: 1 in 500,000

7. Discrepancy between two measures of a section: 20 mm kilometers

8. Probable error or computed distance: 1 in 20,000 to 1 in 50,000

9. Probable error in astronomic azimuth: 2.0 sec

3) *Third Order or Tertiary Triangulation*

The third-order triangulation consists of a number of points fixed within the framework of secondary triangulation, and forms the immediate control for detailed engineering and other surveys. The sizes of the triangles are small and instrument with moderate precision may be used. The specifications for a third-order triangulation are as follows:

1. Average triangle closure: 6 sec
2. Maximum triangle closure: 12 sec
3. Length of base line: 0.5 to 3 km
4. Length of sides of triangles: 1.5 to 10 km
5. Actual error of base: 1 in 75, 0000
6. Probable error of base: 1 in 250,000
7. Discrepancy between two Measures of a section: 25 mm kilometers
8. Probable error or computed distance: 1 in 5,000 to 1 in 20,000
9. Probable error in astronomic Azimuth: 5 sec.

Explain the factors to be considered while selecting base line.

The measurement of base line forms the most important part of the triangulation operations. The base line is laid down with great accuracy of measurement and alignment as it forms the basis for the computations of triangulation system. The length of the base line depends upon the grades of the triangulation. Apart from main base line, several other check bases are also measured at some suitable intervals. In India, ten bases were used, the lengths of the nine bases vary from 6.4 to 7.8 miles and that of the tenth base is 1.7 miles.

Selection of Site for Base Line. Since the accuracy in the measurement of the base line depends upon the site conditions, the following points should be taken into consideration while selecting the site:

1. The site should be fairly level. If, however, the ground is sloping, the slope should be uniform and gentle. Undulating ground should, if possible be avoided.
2. The site should be free from obstructions throughout the whole of the length. The line clearing should be cheap in both labour and compensation.
3. The extremities of the base should be intervisible at ground level.
4. The ground should be reasonably firm and smooth. Water gaps should be few, and if possible not wider than the length of the long wire or tape.
5. The site should suit extension to primary triangulation. This is an important factor since the error in extension is likely to exceed the error in measurement.

In a flat and open country, there is ample choice in the selection of the site and the base may be so selected that it suits the triangulation stations. In rough country, however, the choice is limited and it may sometimes be necessary to select some of the triangulation stations that at suitable for the base line site.

Standard Length. The ultimate standard to which all modern national standards are referred is the international meter established by the Bureau International der Poids at Measures and kept at the Pavilion de Breteuil, Sevres, with copies allotted to various national surveys. The meter is marked on three platinum- iridium bars kept under standard conditions. One great disadvantage of the standard of length that are made of metal are that they are subject to very small secular change in their dimensions. Accordingly, the meter has now been standardized in terms of wavelength of cadmium light.

3.4. Types of Error

Errors of measurement are of three kinds: (i) mistakes, (ii) systematic errors, and (iii) accidental errors.

(i) Mistakes. Mistakes are errors that arise from inattention, inexperience, carelessness and poor judgment or confusion in the mind of the observer. If mistake is undetected, it produces a serious effect on the final result. Hence every value to be recorded in the field must be checked by some independent field observation.

Systematic Error. A systematic error is an error that under the same conditions will always be of the same size and sign. A systematic error always follows some definite mathematical or physical law, and a correction can be determined and applied. Such errors are of constant character and are regarded as positive or negative according as they make the result too great or too small. Their effect is therefore, cumulative. If undetected, systematic errors are very serious. Therefore:

(1) All the surveying equipment must be designed and used so that whenever possible systematic errors will be automatically eliminated and (2) all systematic errors that cannot be surely eliminated by this means must be evaluated and their relationship to the conditions that cause them must be determined. For example, in ordinary levelling, the levelling instrument must first be adjusted so that the line of sight is as nearly horizontal as possible when bubble is cantered. Also the horizontal lengths for back sight and foresight from each instrument position should be kept as nearly equal as possible. In precise levelling, every day, the actual error of the instrument must be determined by careful peg test, the length of each sight is measured by stadia and a correction to the result is applied.

(iii) Accidental Error. Accidental errors are those which remain after mistakes and systematic errors have been eliminated and are caused by a combination of reasons beyond the ability of the observer to control. They tend sometimes in one direction and some times in the other, i.e., they are equally likely to make the apparent result too large or too small.

An accidental error of a single determination is the difference between (1) the true value of the quantity and (2) a determination that is free from mistakes and systematic errors. Accidental error represents limit of precision in the determination of a value. They obey the laws of chance and therefore, must be handled according to the mathematical laws of probability.

The theory of errors that is discussed in this chapter deals only with the accidental errors after all the known errors are eliminated and accounted for.

3.5. Law of Weights

From the method of least squares the following laws of weights are established:

(i) The weight of the arithmetic mean of the measurements of unit weight is equal to the number of observations.

For example, let an angle A be measured six times, the following being the values:

A Weight	A Weight
30°20'8" 1	30°20'10"1
30°20'10"1	30°20'9" 1
30°20'7" 1	30°20'10"1

Arithmetic mean

$$=30°20'+1/6 \ (8'+10' + 7'+10' +9'+10')$$

$$=30°20'9".$$

Weight of arithmetic mean =number of observations =6.

(2) The weight of the weighted arithmetic mean is equal to the sum of the individual weights.

For example, let an angle A be measured six times, the following being the values.

A Weight	A		Weight
30 ° 20'8. 2"	30°20'	10'	3

30°	20'10"	3	30°	20'	9.4"
30°	20'6"	2	30°	20'	10.2"

Sum of weights=2+3 +2 +3 +4 +2=16

Arithmetic mean =30°20'+1/16 (8'X2+10'X3+7'X2 +10'X3+9'X4+10'X2)

=30°20'9". Weight of arithmetic mean =16.

(3) The weight of algebric sum of two or more quantities is equal to the reciprocals of the individual weights.

For Example angle A =30°20'8", Weight 2

B=15°20? 8', Weight 3

Weight of A +B=

(4) If a quantity of given weight is multiplied by a factor, the weight of the result is obtained by dividing its given weight by the square of the factor.

(5) If a quantity of given weight is divided by a factor, the weight of the result is obtained by multiplying its given weight by the square of the factor.

(6) If a equation is multiplied by its own weight, the weight of the resulting equation is equal to the reciprocal of the weight of the equation.

(7) The weight of the equation remains unchanged, if all the signs of the equation are changed or if the equation is added or subtracted from a constant.

3.6. The Law of Accidental Errors and Principles of Least Squares in Survey

The Law of Accidental Errors

Investigations of observations of various types show that accidental errors follow a definite law, the law of probability. This law defines the occurrence of errors and can be expressed in the form of equation which is used to compute the probable value or the probable precision of a quantity. The most important features of accidental errors which usually occur are:

(i) Small errors tend to be more frequent than the large ones; that is they are the most probable.

(ii) Positive and negative errors of the same size happen with equal frequency; that is, they are equally probable.

(iii) Large errors occur infrequently and are impossible.

Principles of Least Squares

It is found from the probability equation that the most probable values of a series of errors arising from observations of equal weight are those for which the sum of the squares is a minimum. The fundamental law of least squares is derived from this. According to the principle of least squares, the most probable value of an observed quantity available from a given set of observations is the one for which the sum of the squares of the residual errors is a minimum. When a quantity is being deduced from a series of observations, the residual errors will be the difference between the adopted value and the several observed values,

Let $V1$, $V2$, $V3$ etc. be the observed values x = most probable value

3.7. Distribution of Error of the Field Measurement

Whenever observations are made in the field, it is always necessary to check for the closing error, if any. The closing error should be distributed to the observed quantities. For examples, the sum of the angles measured at a central angle should be 360°, the error should be distributed to the observed angles after giving proper weight age to the observations. The following rules should be applied for the distribution of errors:

(1) The correction to be applied to an observation is inversely proportional to the weight of the observation.

(2) The correction to be applied to an observation is directly proportional to the square of the probable error.

(3) In case of line of levels, the correction to be applied is proportional to thelength.

The following are the three angles?, ?and y observed at a station P closing the horizon, along with their probable errors of measurement. Determine their corrected values.

Solution

        A= 78 ° 12' 12"   2'

        B = 136 ° 48' 30"   4'

        C= 144 ° 59' 08"   5'

Sum of the three angles = 359 ° 59? 50'

Discrepancy = 10'

Hence each angle is to be increased, and the error of 10' is to be distributed in proportion to the square of the probable error.

Let c1, c2 and c3 be the correction to be applied to the angles?, ? and y respectively.

c1: c2: c3 = $(2)^2$: $(4)^2$: $(5)^2$ = 4 : 16 : 25 _________ (1)

Also, c1 + c2 + c3 = 10" _____________________ (2)

From (1), c2 = 16 /4 c1 = 4c1

And c3 = 25/4 c1

Substituting these values of c2 and c3 in (2), we get c1 +

4c1 + 25/4 c1 = 10'

or c1 (1 + 4 + 25/4)= 10'

c1 = 10 x 4/45 = 0'.89

c2 = 4c1 = 3'.36

And c3 = 25 /4 c1 = 5'.55

Check: c1 + c2 + c3 = 0.'89 + 3'.56 + 5'.55 = 10'

Hence the corrected angles are:

? = 78 ° 12' 12" + 0'.89 = 78 ° 12' 12.89"

? = 136 ° 48' 30" + 3'.56 = 136 ° 48' 33.56"

and y = 144 ° 59' 08" + 5'.55 = 144 ° 59' 13.55"

Sum = 360° 00' 00"+ 00

An angle A was measured by different persons and the following are the values:

Angle Number of measurements.

65 ° 30' 10"Weight   2

65 ° 29' 50"Weight   3

65 " 30' 00"Weight   3

65 ° 30' 20" Weight   4

65 ° 30' 10"Weight   3

Find the most probable value of the angle.

Solution

As stated earlier, the most probable value of an angle is equal to its weighted arithmetic mean.

$$65 ° 30' 10" \times 2 = 131 ° 00' 20"$$
$$65 ° 29' 50" \times 3 = 196 ° 29' 30"$$
$$65 °30' 00" \times 3 = 196 ° 30' 00"$$
$$65 ° 30' 20" \times 4 = 262 ° 01' 20"$$
$$65 °30' 10" \times 3 = 196 ° 30' 30"$$
$$\text{Sum} = 982 ° 31' 40" \quad \text{weight} = 2 + 3 + 3 + 4 + 3 = 15$$

Weighted arithmetic mean = 982°31'40"

= 65°30'6.67"

Hence most probable value of the angle = 65°30'6.67"

The telescope of a theodilite is fitted with stadia wires. It is required to find the most probable values of the constants C and K of tacheometer. The staff was kept vertical at three points in the field and with of sight horizontal the staff intercepts observed was as follows.

Distance of staff   Staff intercept S(m) from tacheometer D(m)

150	1.495
200	2.000
250	2.505

Solution

The distance equation is

$$D = KS + C$$

The observation equations are:

$$150 = 1.495\,K + C$$

$$200 = 2.000\,K + C$$

$$250 = 2.505\,K + C$$

If K and C are the most probable values, then the error of observations are:

$$150 - 1.495\,K - C$$

$$200 - 2.000\,K - C$$

$$250 - 2.505\,K - C$$

By the theory of least squares

$$(150 - 1.495\,K - C)^2 + (200 - 2.000\,K - C)^2 + (250 - 505\,K - C)^2 = minimum \text{---(i)}$$

For normal equation in K,

Differentiating equation (i) w.r.t. K,

$$2(-1.495)(150 - 1.495\,K - C) + 2(-20.)(200 - 2.000\,K - C)$$

$$+2(-2.505)(250 - 505\,K - C) = 0$$

$$208.41667 - 2.085\,K - C = 0 \text{-------- (2)}$$

Normal equation in C

Differentiating equation (i) w.r.t. C,

$$2(-1.0)(150 - 1.495\,K - C) + 2(-1.0)(200 - 2.000\,K - C)$$

$$+2(-1.0)(250 - 505\,K - C) = 0$$

$$200 - 2\,K - C = 0 \text{-------- (3)}$$

On solving Equations (2) and (3)

K = 99.0196

C = 1.9608

The distance equation is:

D = 99.0196 S + 1.9608

The following angles were measured at a station O as to close the horizon

$\angle AOB = 83°42'28".75$	**Weight 3**
$\angle BOC = 102°15'43".26$	**Weight 2**
$\angle COD = 94°38'27".22$	**Weight 4**
$\angle DOA = 79°23'23".77$	**Weight 2**

Adjust the angles by method of Correlates.

Solution

$\angle AOB = 83°42'28".75$	Weight 3
$\angle BOC = 102°15'43".26$	Weight 2
$\angle COD = 94°38'27".22$	Weight 4
$\angle DOA = 79°23'23".77$	Weight 2

____

Sum　　　= 360°00'03".00

Hence the total Correlation E = 360° – (360°00'03")

$$= -3"$$

Let e_1, e_2, e_3 and e_4 be the individual corrections to the four angles respectively. Then by the condition equation, we get

$$e_1 + e_2 + e_3 + e_4 = -3" \underline{\hspace{3cm}} (1)$$

Also, from the least square principle, $\sum(we^2) =$ a minimum

$$3e_1^2 + 2e_2^2 + 4e_3^2 + 2e_4^2 = \text{a minimum} \underline{\hspace{3cm}} (2)$$

Differentiating (1) and (2), we get

$$\delta e_1 + \delta e_2 + \delta e_3 + \delta e_4 = 0 \underline{\hspace{3cm}} (3)$$

BOC = 102°15'43".26 – 0".95　= 102°15'42".31

COD = 94°38'27".22 - 0".47　= 94°38'26".75

DOA = 79°23'23".77 - 0".85　= 79°23'22".82

____

360°00'00".00

The following round of angles were observed from central station to surrounding station of a triangulation survey.

A = 93°43'22"   Weight 3

B = 74°32'39"   Weight 2

C = 101°13'44"   Weight 2

D = 90°29'50"   Weight 3

In addition, one angle $(\overline{A+B})$ was measured separately as combi angle with a mean value of 168°16'06" (wt 2)

Determine the most probable values of the angles A,B,C and D

Solution:

A+B+C+D = 359°59'35"

Total correction E = 360° – (359°59'35")

$$= +25°$$

Similarly, $(\overline{A+B})$ = (A+B)

Hence correction E' = A + B - $(\overline{A+B})$

$$= 168°16'01" - 168°16'06"$$

$$= -5"$$

Let e_1, e_2, e_3, e_4 and e_5 be the individual corrections to A, B, C, D

$(\overline{A+B})$ respectively. Then by the condition equation, we get

$$e_1+e_2+e_3+e_4 = -25" \qquad \text{————————— (1(a))}$$

$$e_5 - e_1-e_2 = -5" \qquad \text{————————— (1(b))}$$

Also, from the least square principle, $\sum(we^2)$ = a minimum

$$3e_1^2 + 2e_2^2 + 2e_3^2 + 3e_4^2 + 2e_5^2 = \text{a minimum} \qquad \text{————————— (2)}$$

Differentiating (1a) (1b) and (2), we get

$$\delta e_1+\delta e_2+\delta e_3+\delta e_4 = 0 \qquad \text{————————— (3a)}$$

$$\delta e_5-\delta e_1+\delta e_2 = 0 \qquad \text{————————— (3b)}$$

$$3e_1\delta e_1+ 2e_2\delta e_2+ 2e_3\delta e_3+ 3e_4\delta e_4 + 2e_5\,\delta e_5 = 0$$

Multiplying equation (3a) by $-\lambda_1$, (3b) by $-\lambda_2$ and adding it to (3), we get

$$\delta e_1 (3e_1- \lambda_{1+} \lambda_2) + \delta e_2 (2e_2- \lambda_{1+} \lambda_2) + \delta e_3 (2e_3- \lambda_1) + \delta e_4 (3e_4- \lambda_1) + \delta e_5 (- \lambda_2 + 3e_5) = 0$$
$$\text{————(5)}$$

Since the coefficients of δe_1, δe_2, δe_3, δe_4 etc must vanish independently we have,

$$-\lambda_{1+}\lambda_2 + 3e_1 = 0 \quad \text{or} \quad e_1 = \frac{\lambda 1}{3} - \frac{\lambda 2}{3}$$

$$-\lambda_{1+}\lambda_2 + 2e_2 = 0 \quad e_2 = \frac{\lambda 1}{2} - \frac{\lambda 2}{2}$$

$$-\lambda_2 + 2e_3 = 0 \quad e_3 = \frac{\lambda_1}{2} \underline{\hspace{2cm}} (6)$$

$$-\lambda_1 + 3e_4 = 0 \quad e_4 = \frac{\lambda_1}{3}$$

$$-\lambda_2 + 2e_5 = 0 \quad e_1 = -\frac{\lambda_2}{2}$$

Substituting these values of e_1, e_2, e_3, e_4 and e_5 in equations (1a) and (1b)

$$\frac{\lambda_1}{3} - \frac{\lambda_2}{3} + \frac{\lambda_1}{2} - \frac{\lambda_2}{2} + \frac{\lambda_1}{2} + \frac{\lambda_1}{3} = 25 \text{ from (1a)}$$

Or $\quad 5\frac{\lambda_1}{3} - \frac{5}{6}\lambda_2 = 25$

$$\frac{\lambda_1}{3} - \frac{5}{6}\lambda_2 = 5 \underline{\hspace{2cm}} (I)$$

$$\frac{\lambda_2}{2} - \frac{\lambda_1}{3} + \frac{\lambda_2}{3} - \frac{\lambda_1}{2} + \frac{\lambda_2}{32} = -5 \text{ from (b)}$$

$$4\frac{\lambda_2}{3} - \frac{5}{6}\lambda_1 = -5 \underline{\hspace{2cm}} (II)$$

Solving (I) and (II) simultaneously, we get

$$\lambda_1 = +\frac{210}{11}$$

$$\lambda_2 = +\frac{90}{11}$$

Hence

$$e_1 = \frac{1}{3}\cdot\frac{210}{11} \; \frac{1}{3}\cdot\frac{90}{11} = \frac{40''}{11} = 3''.64$$

$$e_2 = \frac{1}{2}\cdot\frac{210}{11} - \frac{1}{2}\cdot\frac{90}{11} = \frac{60''}{11} = 5''.45$$

$$e_3 = \frac{1}{2}\cdot\frac{210}{11} = \frac{105''}{11} = 9''.55$$

$$e_4 = \frac{1}{3}\cdot\frac{210}{11} = \frac{70''}{11} = 6''.36$$

$$\text{Total} = +25''.00$$

Also,

$$e_5 = \frac{1}{2}\cdot\frac{90}{11} + 4''09$$

Hence the corrected angles are

$$A = 93°43'22'' + 3''.64 \quad = 93°43'25''.64$$

$$B = 74°32'39'' + 5''.45 \quad = 74°32'44''.45$$

$$C = 101°13'44'' + 9''.55 \quad = 101°13'53''.55$$

$$D = 90°29'50'' + 6''.36 \quad = 90°29'56''.36$$

$$\text{Sum} = 360°00'00''.00$$

MODEL SHORT QUESTIONS AND ANSWERS

CONTROL SURVEYING AND ADJUSTMENT

1. **Define true error and residual error? (April/May 2018)(April/May 2017) (Nov/Dec 2014)**

   - A true error is the difference between the true value of a quantity and its observed value.

   - A residual error is the difference between the most probable value of a quantity and its observed value.

2. **What is reciprocal observation? (April/May 2018)**

   - One of two measurements made as a pair to reduce the size of some systematic error in the individual measurement. In particular, one or a pair of measurement taken forward and backward at the end of the line. Example-Reciprocal leveling.

3. **Give the specification of first order triangulation. (April/May 2018)**

   - The specification of first order triangulation is of the highest order and is employed either to determine the earth figure or to finish the most precise control points to which secondary triangulation may be connected.

4. **Define triangulation. (Nov/Dec 2017) (April/May 2015)**

   - Triangulation is the process of determining the location of a point by forming triangles to it from known points.

5. **What is weight of observation? (Nov/Dec 2017) (Nov/Dec 2014)**

   - Weight of an observation is a measure of its relative worth which may be indicated by a number. Thus if a certain observation is said to have weight age five, it is meant to say that it is five times as much as an observation of weigh tone.

6. **What is satellite station? (Nov/Dec 2017) (Nov/Dec 2012)**

   - In order to form to well-conditioned triangles of triangulation and also to have better visibility objects such as church spirals, towers of temples, flag poles, etc are selected. But the instrument cannot be set up over these true stations for the measurement of angles. In such cases, a subsidiary station called as satellite station or eccentric station or false station, is selected as near as possible to the true station.

7. **What is meant by phase of signals? (April/May 2017) (Nov/Dec 2016)**

   - When a cylindrical signal is partly illuminated and partly in shade, the observer sees only the illuminated portion and bisects it. The error of bisection thus introduced is called phase. It is apparent displacement of the centre of the signal.

8. **What are the classifications of errors? (April/May 2017) (Nov/Dec 2012)**

- Mistakes

- Systematic errors

- Accidental errors

9. **State the principle of least squares. (Nov/Dec 2016) (April/May 2015)**

- "In observations of equal precision the most probable values of the observed quantities are those that render the sum of the squares of the residual errors a minimum"

10. **What is parallax? How it can be eliminated. (Nov/Dec2016)**

Parallax is a displacement or difference in the apparent position of an object viewed along two different lines of sight, and is measured by the angle or semi-angle of inclination between those two lines.

11. **What is true and most probable value? (April/May 2015)**

- The true value of a quantity is the value which is absolutely free from all the errors. It is indeterminate since the true error is never known.

- Most probable value of a quantity is the value which is more likely to be the true value than any other value.

12. **What is the application Gale's table? (April/May 2015)**

- Traverse computations are usually done in tabular form. A more common tabular form is "Gales Traverse Table".

- The reduced bearings of all the lines are computed based on the whole circle bearing of the lines.

- The latitudes and departures of all the lines are computed.

13. **Define Tacheometry**

Tacheometryis a branch of angular surveying in which the horizontal and vertical distances (or) points are obtained by optional means as opposed to the ordinary slower process of measurements by chain (or)tape.

14. **Define Tacheometer**

It is an ordinary transit theodolite fitted with an extra lens called analytic lens.

The purpose of fitting the analytic lens is to reduce the additive constant to zero.

15. Define Analytic lens

Analytic lens is an additional lens placed between the diaphragm and the objective at a fixed distance from the objective. This lens will be fitted in ordinary transit theodolite. After fitting this additional lens the telescope is called as external focusing analytic telescope. The purpose of fitting the analytic lens is to reduce the additive constant to zero.

16. Define Substance bar

A Substance bar is manufactured by Mr. Kern. The length of the substance bar is 2m (6ft) for measurement of comparatively short distance in a traverse. A Substance bar may be used as a substance base. The length of the bar is made equal to the distance between the two targets.

17. What are the merits and demerits of movable hair method?

Merits

Long sights can be taken with greater accuracy than stadia method. The error obtained is minimum.

Demerits

The computations are not quicker. Careful observation is essential.

18. What is the objective of geodetic surveying?

Geodetic surveying is also called as trigonometrical surveying which deals with long distances and larger areas. The objects of geodetic survey is to establish absolute and relative positions of a number of widely separated points on the earth`s surface.

19. Distinguish between triangulation and trilateration.

Triangulation is a survey by which position of several stations are fixed very accurately on the surface of the earth at large intervals which serve as basis or reference points.

Trilateration is based on the principle that a triangle can be solved by knowing its three sides; Instruments like geodimeter and tellurometer are employed. In geodetic survey this method is extensively used and the accuracy of the results is comparable to that of triangulation.

20. What are the different classifications of triangulation system?

Classification of a triangulation system is based on the accuracy with which the length and angle of a line of a triangulation are determined. The following are the classification based on the order of grades:

    (i) First order or primary triangulation.

(ii) Second order or secondary triangulation.

(iii) Third order or tertiary triangulation.

21. What is meant by phase of a signal?

When a cylindrical signal is partly illuminated and partly in shade, the observer sees only the illuminated portion and bisects it. The error of bisection thus introduced is called phase. It is the apparent displacement of the centre of the signal.

22. List out corrections for tape.

(i) **Correction for temperature**: This correction may be positive or negative depending on the temperature at the time of measurement (T_m) and the standard temperature (T_0).

(ii) **Correction of pull**: Correction for pull will be positive or negative depending on the applied pull (P_m) is greater or less than standard pull (P_0).

(iii) **Correction for slope.**

(iv) **Correction for sag.**

23. What is a base net?

Some site conditions may not be favorable to get the required length of a base line. In such a situation a short base line is selected and the same is then extended. Such group of triangles which are meant for extending the base is known as base net.

24. What are the different kinds of benchmarks?

GTS bench marks: These bench-marks are established all over the country at large interval by the survey of India Department.

Permanent bench marks: These are fixed points or marks set up by different Govt. Department. The reduced levels of these points are determined with references to GTS bench marks.

Arbitrary bench marks: When reduced levels of some fixed points are assumed they are called as arbitrary bench marks. These are adopted in small survey operations.

Temporary bench marks: When bench marks are set up temporarily at the end of a day`s work, they are referred to as temporary bench-marks.

25. What do you understand by the term Traversing?

A traverse is a multi-sided figure consisting of a series of connected lines. The lengths are measured by chain or tape and the directions are identified by angle measuring instruments.

Closed traverse is one in which the last survey line is joined back to the first station point forming a polygon. Open traverse, called as unclosed transverse, when it does not return back to the first station point and form a closed polygon.

26. What is called axis signal correction?

At the stations, signals are erected at different heights. The signals may or may not be of the same height as that of the instrument. If the height of the signal is not the same as that of the height of the instrument axis above the station, a correction known as the axis signal correction or eye and object correction is to be applied.

27. Define baseline.

Base line forms the basis for the entire computations of triangulation system. The length of base line to be adopted depends on the magnitude of triangulation work, i.e., the grade of the triangulation.

28. Define Satellite Station.

In order to form well conditioned triangles of triangulation and also to have better visibility objects such as church spirals, towers of temples, flag poles, etc are selected. But the instrument cannot be setup over these true stations for the measurements of angles and a subsidiary station called as satellite station or eccentric station or false station.

29. Define reduction to centre.

Angles taken from satellite are corrected and reduced to what they would have been if the true station was occupied. This operation of applying corrections to the observed angles due to the eccentricity of the station is termed as Reduction to centre.

30. What are the Methods used to measure baseline.

The field work for the base line measurements is carried out by two parties,

    (i) Setting out party

This party consists of two surveyors and a number of porters. The duty of the porters is to place the measuring tripods. At correct intervals, in alignment in advance.

    (ii) Measuring Party

This party consists of two observers, recorder, leveler and staff man for actual measurements.

31. Mention the Two types of trigonometrical leveling.

Trigonometrical levelling may be grouped as,

    (i)   Observations to find small elevations and short distances.

    (ii) Observations to find higher elevations and large distances.

32. Define Control Surveying.

- A survey that provides coordinates (H&V) of points to which supplementary survey are adjusted.
- A survey which is performed to achieve higher than normal accuracies.

33. Explain the terms true error and most probable error.

A true error is the difference between the true value of a quality and its observed value. Most probable error is defined as that quantity which is added to, or subtracted from, the most probable value which fixes the limits. By these limits there is an even chance the true of the measured quantity may lie.

34. Distinguish between true error and residual error.

- A true error is the different between the true value of a quantity and its observed value.
- A residual error is the difference between the most probable value of a quantity and its observed value.

35. State the principle of method of least squares.

In observations of equal precision the most probable values of the observed quantities are those that render the sum of the squares of the residual errors a minimum.

36. What are the kinds of errors possible in survey work?

Error made on an observation may be due to some reason. Error may be classified in a more general form as (i) mistakes, (ii) systematic error and (iii) accidental error. Value of an error is also assigned as true, most probable and residual.

37. What is the weight of an observation?

Weight of an observation is a measure of its relative worth which may be indicated by a number. Thus if a certain observation is said to have weight age 5, it is meant to say that it is five times as much as an observation of weight 1.

38. How are normal equations formed in theory of errors?

A normal equation is an equation of condition by means of which the most probable value of any unknown quantity may be determined corresponding to a set of values assigned to other unknown quantities. Therefore normal equations have to be formed for each of the unknown, to determine their values.

39. What is method of correlates?

- Correlates are the unknown multiples or independent constants employed for finding the most portable values of unknowns.

- In this method of correlates all the condition equations are collected. One more equation of condition, i.e., the sum of the residual errors should be minimum, is added.

40. Define normal equation.

A normal equation is an equation of condition by means of which the most probable value of any unknown quantity may be determined.

41. Distinguish between the observed value and the most probable value of a quantity.

- An observed value is the numerical value of a measured quantity. This may be a direct observation or an indirect observation.

- Most probable value of a quantity is the value which is more likely to be the true value that any other value. It is the one which is deduced from the several measurements or which it is based.

42. Distinguish between true value and most probable value.

- True value of a quantity is the value which is absolutely free from all the errors. It is indeterminate since the true error is never known.

- Most probable value of a quantity is the value which is more likely to be the true value than any other value.

43. What do you mean by figure adjustments in triangulation?

Figure adjustment is the determination of the most probable values of the angles involved in any geometric figure so as to fulfill, the geometric requirements. It invariably involves one or more conditional equations. Conditional equations may be framed by the method of normal equation or by the method of correlated.

44. What is single angle adjustment?

In general single angle is measured several times. Corrections to be applied are inversely proportional to the weight and directly proportional to the square of probable errors.

In case of equal weighted measurements the most probable value is equal to the arithmetic mean of the observations. In case of unequally weighted observations, the most probable value of the angle is equal to the weighted arithmetic mean of the observed angles.

45. Why figure adjustment is made?

Figure adjustment is needed so as fulfill the geometric conditions of nay geometrical figures. The figure adjustment, therefore involves one or more condition equations.

46. State Gauss`s rule.

It is applied when the weights of the observations are not directly known. If the residual error of each observation is known the weights can be calculated by Gauss`s rule.

47. What is called spherical excess?

In a spherical triangle the sum of the three angles of the triangles always exceeds 180^0 by an amount known as spherical excess. Spherical excess depends on the area of a triangle. It may be taken approximately equivalent to 1" for every 196.75sq.km.

48. Explain level net.

A level net is an interconnecting net work of level circuits formed by level lines interconnecting three or more bench marks. In adjusting a level net, the method of least squares may be adopted.

49. Define systematic error.

A systematic error is one which occurs under a given condition and be of same size and sign. Such errors generally add up to make the results too great or too small. That is why it is also called as cumulative error. Such error follows some definite mathematical or physical law and hence a correction can be applied.

50. Define Accidental error.

It occurs by a combination of reasons beyond the ability of the observer to control. They sometimes occur in one direction and some times in the other side. Thus they are likely to make the apparent result too large or too small.

51. What is mean by single angle adjustment?

In case of equal weighted measurements the most probable value is equal to the arithmetic mean of the observations. In case of unequally we8ighted observations, the most probable value of the angle is equal to the weighted arithmetic mean of the observed angles.

52. Define Figure adjustments?

It is the determination of the most probable values of the angles involved in any geometrical figure so as to fulfill the geometric requirements. Conditioned equations may be framed by the method of normal equation or by the method of correlates.

MODEL DETAIL QUESTIONS

1) What is meant by triangulation adjustment? Explain the different conditions and cases with sketches. **(April/May 2018)**

2) A traverse ABCD was to be run but due to an obstruction between the stations A and B it was not possible to measure the length and direction of the line AB. The following data could only be obtained.

line	Length(m)	R.B
AD	44.5	N50^{0}20'E
DC	67.0	S69^{0}45'E
CB	61.3	S30^{0}10'W

Determine the length and the direction of BA. Also determine the perpendicular distance of C from AB. **(April/May 2018)**

3) Find the most probable values of the angles A,B and C from the following observation. **(April/May 2018) (April/May 2015)**

$$A=68^0 12'36''$$

$$B=53^0 46'12''$$

$$C=58^0 01'16''$$

4) Find the most probable values of the angles A,B and C from the following observation. **(Nov/Dec 2017) (April/May 2017)**

$$A=75^0 32'42.3''$$

$$B=55^0 09'53.2''$$

$$C= 108^0 09'28.8''$$

$$A+B= 130^0 42'4.6''$$

$$B+C= 163^0 19'22.5''$$

$$A+B+C=232^0 52'9.8''$$

5) Ajust the following angles closing the horizon: **(Nov/Dec 2017)**

$$A= 110^0 20'48''wt.4$$

$$B= 92^0 30'12'' wt.1$$

$$C= 56^0 12'00'' wt.2$$

$$D= 100^0 57'04'' wt.3$$

6) The angles of the triangles ABC were recorded as Angle POQ = $83^0 42'28.75''$ weight 3; Angle QOR = $102^0 15'43.26''$ weight 2; Angle ROS = $94^0 38'27.22''$ weight 4; Angle SOP = $79^0 23'23.77''$ weight 2; Give the corrected values of the angles. **(April/May 2017) (Nov/Dec 2014)**

7) Find the most probable values of the angles A, B and C from the following observation. **(April/May 2017)**

$$A = 32^0 15'3.62'' \qquad wt\text{-}2$$
$$B = 40^0 16'18.4'' \qquad wt\text{-}1$$
$$C = 35^0 12'26.6'' \qquad wt\text{-}1$$
$$A+B = 72^0 31'50.2'' \qquad wt\text{-}1$$
$$A+B+C = 107^0 44'25.5'' \qquad wt\text{-}2$$

8) Find the most probable value of angles A,B and C of triangle ABC from the following observation equations: **(Nov/Dec 2016) (April/May 2015)**

$$A = 9^0 48'36.6'' \ wt.2$$
$$B = 54^0 37'48.3'' \ wt.3$$
$$A+B = 104^0 26'28.5'' \ wt.4$$

9) .

(a) What are the laws of random errors? **(Nov/Dec 2016) (Nov/Dec 2014)**

(b) The angles of the triangles ABC were recorded as A= $77^0 14'20''$ weight 4; B= $49^0 40'35''$ weight 3; C= $53^0 04'52''$ weight 2; Give the corrected values of the angles.

10) .

(a) Explain the order of triangulation and its specification.

(b) Find the R.L of Q from the following observations;

    Horizontal distance between P and Q = 9290m

    Angle of elevation from P to Q = $2^0 06'18''$

    Height of signal at Q=3.96m

    Height of instrument at P= 1.25m

    Coefficient of refraction= 0.07

    R sin 1'' = 30.88m

    R.L of P= 396.58m **(Nov/Dec 2015) (Nov/Dec 2013)**

Unit IV

4. Advanced Topics in Surveying

Hydrographic Surveying – Tides – MSL – Sounding methods – Three point problem – Strength of fix – astronomical Surveying – Field observations and determination of Azimuth by altitude and hour angle methods – Astronomical terms and definitions - Motion of sun and stars – Celestial coordinate systems – different time systems – Nautical Almanac – Apparent altitude and corrections – Field observations and determination of time, longitude, latitude and azimuth by altitude and hour angle method

4.1. Introduction

Photogrammetry - Introduction

- Terrestial and aerial Photographs
- Stereoscopy
- Parallax
- Electromagnetic distance measurement
- Carrier waves
- Principles - Instruments
- Trilateration

Hydrographic Surveying

- Tides
- MSL
- Sounding methods
- Location of soundings and methods
- Three point problem
- Strength of fix
- Sextants and station pointer
- River surveys
- Measurement of current and discharge

Cartography

- Cartographic concepts and techniques
- Cadastral surveying
- Definition
- Uses

- Legal values

- Scales and accuracies.

4.2. Photogrammetric Surveying

Photogram metric surveying or photogrammetry is the science and art of obtaining accurate measurements by use of photographs, for various purposes such as the construction of planimetric and topographic maps, classification of soils, interpretation of geology, acquisition of military intelligence and the preparation of composite pictures of the ground.

4.2.1. Principles Behind Terrestrial Photogrammetry

The principle of terrestrial photogrammetry was improved upon and perfected by Capt. Deville, then Surveyor General of Canada in 1888. In terrestrial photogrammetry, photographs are taken with the camera supported on the ground. The photographs are taken by means of a photo theodolite which is a combination of a camera and a theodolite. Maps are then compiled from the photographs.

4.2.2. Photogrammetric Surveying

Photogram metric surveying or photogrammetry is the science and art of obtaining accurate measurements by use of photographs, for various purposes such as the construction of planimetric and topographic maps, classification of soils, interpretation of geology, acquisition of military intelligence and the preparation of composite pictures of the ground. The photographs are taken either from the air or from station on the ground. Terrestrial photogrammetry is that Brach of photogrammetry wherein photographs are taken from a fixed position on or near the ground. Aerial photogrammetry is that branch of photogrammetry wherein the photographs are taken by a camera mounted in an aircraft flying over the area.

Mapping from aerial photographs is the best mapping procedures yet developed for large projects, and are invaluable for military intelligence. The major users of aerial mapping methods are the civilian and military mapping agencies of the Government.

The conception of using photographs for purposes of measurement appears to have originated with the experiments of Aime Laussedat of the Corps of the French Army, who in 1851 produced the first measuring camera. He developed the mathematical analysis of photographs as perspective projections, thereby increasing their application to topography. Aerial photography from balloons probably began about 1858. Almost concurrently (1858), but independently of Laussedat, Meydenbauer in Germany carried out the first experiments in making critical measurements of architectural details by the intersection method in the basis of two photographs of the building. The ground photography was perfected in Canada by Capt. Deville, then Surveyor General of Canada in 1888. In Germany, most of the progress on the theoretical side was due to Hauck.

In 1901, Pulfrich in Jena introduced the stereoscopic principle of measurement and designed the stereo comparator. The stereoaitograph was designed (1909) at the Zeiss workshops in Jena, and this opened a wide field of practical application. Scheimpflug, an Australian captain, developed the idea of double projector in 1898. He originated the theory of perspective transformation and incorporated its principles in the photoperspecto graph. He also gave the idea of radial triangulation. His work paved the way for the development of aerial surveying and aerial photogrammetry.

In 1875, Oscar Messter built the first aerial camera in Germany and J.W.Bagloy and A.Brock produced the first aerial cameras in U.S.A. In 1923, Bauersfeld designed the Zeiss stereoplanigraph. The optical industries of Germany, Switzerland, Italy and France, and later also those of the U.S.A and U.S.S.R. took up the manufacture and constant further development of the cameras and plotting instruments. In World War II, both the sides made extensive use of aerial photographs for their military operations. World War II gave rise to new developments of aerial photography techniques, such as the application of radio control to photoflight navigation, the new wide-angle lenses and devices to achieve true vertical photographs.

4.2.3. Methods Employed in Locating Soundings

The soundings are located with reference to the shore traverse by observations made

(i) Entirely from the Boat, (ii) Entirely from the Shore or (iii) From both.

The following are the methods of location

1. By cross rope.
2. By range and time intervals.
3. By range and one angle from the shore.
4. By range and one angle from the boat.
5. By two angles from the shore.
6. By two angles from the boat.
7. By one angle from shore and one from boat.
8. By intersecting ranges.
9. By tacheometry. Range.

A range or range line is the line on which soundings are taken. They are, in general, laid perpendicular to the shore line and parallel to each other if the shore is straight or are arranged radiating from a prominent object when the shore line is very irregular.

Shore Signals

Each range line is marked by means of signals erected at two points on it at a considerable distance apart. Signals can be constructed in a variety of ways. They should be readily seen and easily distinguished from each other. The most satisfactory and economic type of signal is a wooden tripod structure dressed with white and coloured signal of cloth. The position of the signals should be located very accurately since all the soundings are to be located with reference to these signals.

Location by Cross-Rope

This is the most accurate method of locating the soundings and may be used for rivers, narrow lakes and for harbours. It is also used to determine the quantity of materials removed by dredging the soundings being taken before and after the dredging work is done. A single wire or rope is stretched across the channel etc. and is marked by metal tags at appropriate known distance along the wire from a reference point or zero station on shore. The soundings are then taken by a weighted pole. The position of the pole during a sounding is given by the graduated rope or line.

In another method, specially used for harbours etc., a reel boat is used to stretch the rope. The zero end of the rope is attached to a spike or any other attachment on one shore. The rope is would on a drum on the reel boat. The reel boat is then rowed across the line of sounding,

Thus unwinding the rope as it proceeds. When the reel boat reaches the other shore, its anchor is taken ashore and the rope is wound as tightly as possible. If anchoring is not possible, the reel is taken ashore and spiked down. Another boat, known as the sounding boat, then starts from the previous shore and soundings are taken against each tag of the rope. At the end of the soundings along that line, the reel boat is rowed back along the line thus winding in the rope. The work thus proceeds.

Location by Range and Time Intervals

In this method, the boat is kept in range with the two signals on the shore and is rowed along it at constant speed. Soundings are taken at different time intervals. Knowing the constant speed and the total time elapsed at the instant of sounding, the distance of the total point can be known along the range. The method is used when the width of channel is small and when great degree of accuracy is not required. However, the method is used in conjunction with other methods, in which case the first and the last soundings along a range are located by angles from the shore and the intermediate soundings are located by interpolation according to time intervals.

Location by Range and One Angle from the Shore

In this method, the boat is ranged in line with the two shore signals and rowed along the ranges. The point where sounding is taken is fixed on the range by observation of the angle from the shore. As the boat proceeds along the shore, other soundings are also fixed by the observations of angles from the shore. Thus B is the instrument station, A1 A2 is the range along which the boat is rowed and? 1, 2, 3 etc., are the angles measured at B from points 1, 2, 3 etc. The method is very accurate and very convenient for plotting. However, if the angle at the sounding point (say angle) is less than 30°, the fix becomes poor. The nearer the intersection angle is to a right angle, the better. If the angle diminishes to about 30° a new instrument station must be chosen. The only defect of the method is that the surveyor does not have an immediate control in all the observation. If all the points are to be fixed by angular observations from the shore, a note-keeper will also be required along with the instrument man at shore since the observations and the recordings are to be done rapidly. Generally, the first and last soundings and every tenth sounding are fixed by angular observations and the intermediate points are fixed by time intervals. Thus the points with round mark are fixed by angular observations from the shore and the points with cross marks are fixed by time intervals. The arrows show the course of the boat, seaward and shoreward on alternate sections.

To fix a point by observations from the shore, the instrument man at B orients his line of sight towards a shore signal or any other prominent point (known on the plan) when the reading is zero. He then directs the telescope towards the leadsman or the bow of the boat, and is kept continually pointing towards the boat as it moves. The surveyor on the boat holds a flag for a few seconds and on the fall of the flag, the sounding and the angle are observed simultaneously.

The angles are generally observed to the nearest 5 minutes. The time at which the flag falls is also recorded both by the instrument man as well as on the boat. In order to avoid acute intersections, the lines of soundings are previously drawn on the plan and suitable instrument stations are selected.

Location by Range and One Angle from the Boat

The method is exactly similar to the previous one except that the angular fix is made by angular observation from the boat. The boat is kept in range with the two shore signals and is rowed along it. At the instant the sounding is taken, the angle, subtended at the point between the range and some prominent point B on the sore is measured with the help of sextant. The telescope is directed on the range signals, and the side object is brought into coincidence at the instant the sounding is taken. The accuracy and ease of plotting is the same as obtained in the previous method. Generally, the first and the last soundings, and some of the intermediate soundings are located by angular observations and the rest of the soundings are located by time intervals.

As compared to the previous methods, this method has the following advantages:

1. Since all the observations are taken from the boat, the surveyor has better control over the operations.
2. The mistakes in booking are reduced since the recorder books the readings directly as they are measured.
3. On important fixes, check may be obtained by measuring a second angle towards some other signal on the shore.
4. To obtain good intersections throughout; different shore objects may be used for reference to measure the angles.

Location by Two Angles from the Shore

In this method, a point is fixed independent of the range by angular observations from two points on the shore. The method is generally used to locate some isolated points. If this method is used on an extensive survey, the boat should be run on a series of approximate ranges. Two instruments and two instrument men are required. The position of instrument is selected in such a way that a strong fix is obtained. New instrument stations should be chosen when the intersection angle falls below $3°$. Thus A and B are the two instrument stations. The distance d between them is very accurately measured. The instrument stations A and B are precisely connected to the ground traverse or triangulation, and their positions on plan are known. With both the plates clamped to zero, the instrument man at A bisects B; similarly with both the plates clamped to zero, the instrument man at B bisects A. Both the instrument men then direct the line of sight of the telescope towards the leadsman and continuously follow it as the boat moves.

The surveyor on the boat holds a flag for a few seconds, and on the fall of the flag the sounding and the angles are observed simultaneously. The co-ordinates of the position P of the sounding may be computed from the relations:

The method has got the following advantages:

1. The preliminary work of setting out and erecting range signals is eliminated.

2. It is useful when there are strong currents due to which it is difficult to row the boat along the range line.

The method is, however, laborious and requires two instruments and two instrument. Location by Two Angles from the Boat.

In this method, the position of the boat can be located by the solution of the three- point problem by observing the two angles subtended at the boat by three suitable shore objects of known position. The three-shore points should be well-defined and clearly visible. Prominent natural objects such as church spire, lighthouse, flagstaff, buoys etc., are selected for this purpose. If such points are not available, range poles or shore signals may be taken. Thus A, B and C are the shore objects and P is the position of the boat from which the angles are measured. Both the angles should be observed simultaneously with the help of two sextants, at the instant the sounding is taken. If both the angles are observed by surveyor alone, very little time should be lost in taking the observation. The angles on the circle are read afterwards. The method is used to take the soundings at isolated points. The surveyor has better control on the operations since the survey party is concentrated in one boat. If sufficient number of prominent points are available on the shore, preliminary work o setting out and erecting range signals is eliminated. The position of the boat is located by the solution of the three point problem either analytically or graphically.

Location by One Angle from the Shore and the other from the Boat

This method is the combination of methods 5 and 6 described above and is used to locate the isolated points where soundings are taken. Two points A and B are chosen on the shore, one of the points (say A) is the instrument station where a theodolite is set up, and the other (say B) is a shore signal or any other prominent object. At the instant the sounding is taken at P, the angle at A is measured with the help of a sextant. Knowing the distance d between the two points A and B by ground survey, the position of P can be located by calculating the two co-ordinates x and y.

Location by Intersecting Ranges

This method is used when it is required to determine by periodical sounding at the same points, the rate at which silting or scouring is taking place. This is very essential on the harbours and reservoirs. The position of sounding is located by the intersection of two ranges, thus completely avoiding the angular observations. Suitable signals are erected at the shore. The boat is rowed along a range perpendicular to the shore and soundings are taken at the points in which inclined ranges intersect the range. However, in order to avoid the confusion, a definite system of flagging the range poles is necessary. The position of the range poles is determined very accurately by ground survey.

Location by Tacheometric Observations

The method is very much useful in smooth waters. The position of the boat is located by tacheometric observations from the shore on a staff kept vertically on the boat. Observing the staff intercept s at the instant the sounding is taken, the horizontal distance between the instrument stations and the boat is calculated by the direction of the boat (P) is established by observing the angle at the instrument station B with reference to any prominent object A The transit station should be near the water level so that there will be no need to read vertical angles. The method is unsuitable when soundings are taken far from shore.

4.2.4. Tides and its Types and Formation

All celestial bodies exert a gravitational force on each other. These forces of attraction between earth and other celestial bodies (mainly moon and sun) cause periodical variations in the level of a water surface, commonly known as tides. There are several theories about the tides but none adequately explains all the phenomenon of tides. However, the commonly used theory is after Newton, and is known as the equilibrium theory. According to this theory, a force of attraction exists between two celestial bodies, acting in the straight line joining the centre of masses of the two bodies, and the magnitude of this force is proportional to the product of the masses of the bodies and is inversely proportional to the square of the distance between them. We shall apply this theory to the tides produced on earth due to the force of attraction between earth and moon. However, the following assumptions are made in the equilibrium theory:

1. The earth is covered all round by an ocean of uniform depth.

The ocean is capable of assuming instantaneously the equilibrium, required by the tide producing forces. This is possible if we neglect (i) inertia of water, (ii) viscosity of water, and (iii) force of attraction between parts of itself.

1. The Lunar Tides shows the earth and the moon, with their centres of masses O_1 and O_2 respectively. Since moon is very near to the earth, it is the major tide producing force. To start with, we will ignore the daily rotation of the earth on its axis. Both earth and moon attract each other, and the force of attraction would act along O_1, O_2. Let O be the common centre of gravity of earth and moon. The earth and moon revolve monthly about O, and due to this revolution their separate positions are maintained. The distribution of force is not uniform, but it is more for the points facing the moon and less for remote points. Due to the revolution of earth about the common centre of gravity O, centrifugal force of uniform intensity is exerted on all the particles of the earth. The direction of this centrifugal force is parallel to O_1O_2 and acts outward. Thus, the total force of attraction due to moon is counter-balanced by the total centrifugal force, and the earth maintains its position relative to the moon. However, since the fore of attraction is not uniform, the resultant force will very all along. The resultant forces are the tide producing forces. Assuming that water has no inertia and viscosity, the ocean enveloping the earth's surface will adjust itself to the unbalanced resultant forces, giving rise to the equilibrium. Thus, there are two lunar tides at A and B, and two low water positions at C and D. The tide at A is called the superior lunar tide or tide of moon's upper transit, while tide at B is called inferior or antilunar tide.

Now let us consider the earth's rotation on its axis. Assuming the moon to remain stationary, the major axis of lunar tidal equilibrium figure would maintain a constant position. Due to rotation of earth about its axis from west to east, once in 24 hours, point A would occupy successive position C, B and D at intervals of 6 h. Thus, point A would experience regular variation in the level of water. It will experience high water (tide) at intervals of 12 h and low water midway between. This interval of 6 h variation is true only if moon is assumed stationary. However, in a lunation of 29.53 days the moon makes one revolution relative to sun from the new moon to new moon. This revolution is in the same direction as the diurnal rotation of earth, and hence there are 29.53 transits of moon across a meridian in 29.53 mean solar days. This is on the assumption that the moon does this revolution in a plane passing through the equator. Thus, the interval between successive transits of moon or any meridian will be 24 h, 50.5 m. Thus, the average interval between successive high waters would be about 12 h 25 m. The interval of 24 h 50.5 m between two successive transits of moon over a meridian is called the tidal day.

2. The Solar Tides

The phenomenon of production of tides due to force of attraction between earth and sun is similar to the lunar tides. Thus, there will be superior solar tide and an inferior or anti-solar tide. However, sun is at a large distance from the earth and hence the tide producing force due to sun is much less.

Solar tide = 0.458 Lunar tide. Combined effect: Spring and neap tides

Solar tide = 0.458 Lunar tide.

Above equation shows that the solar tide force is less than half the lunar tide force. However, their combined effect is important, especially at the new moon when both the sun and moon have the same celestial longitude, they cross a meridian at the same instant.

Assuming that both the sun and moon lie in the same horizontal plane passing through the equator, the effects of both the tides are added, giving rise to maximum or spring tide of new moon. The term 'spring' does not refer to the season, but to the springing or waxing of the moon. After the new moon, the moon falls behind the sun and crosses each meridian 50 minutes later each day. In after 7 days, the difference between longitude of the moon and that of sun becomes 90°, and the moon is in quadrature. The crest of moon tide coincides with the trough of the solar tide, giving rise to the neap tide of the first quarter. During the neap tide, the high water level is below the average while the low water level is above the average. After about 15 days of the start of lunation, when full moon occurs, the difference between moon's longitude and of sun's longitude is 180°, and the moon is in opposition.

However, the crests of both the tides coincide, giving rise to spring tide of full moon. In about 22 days after the start of lunation, the difference in longitudes of the moon and the sun becomes 270° and neap tide of third quarter is formed. Finally, when the moon reaches to its new moon position, after about 29 days of the previous new moon, both of them have the same celestial longitude and the spring tide of new moon is again formed making the beginning of another cycle of spring and neap tides.

3. Other Effects

The length of the tidal day, assumed to be 24 hours and 50.5 minutes is not constant because of (i) varying relative positions of the sun and moon, (ii) relative attraction of the sun and moon, (iii) ellipticity of the orbit of the moon (assumed circular earlier) and earth, (v) declination (or deviation from the plane of equator) of the sun and the moon, (v) effects of the land masses and (vi) deviation of the shape of the earth from the spheroid. Due to these, the high water at a place may not occur exactly at the moon's upper or lower transit. The effect of varying relative positions of the sun and moon gives rise to what are known as priming of tide and lagging of tide.

At the new moon position, the crest of the composite tide is under the moon and normal tide is formed. For the positions of the moon between new moon and first quarter, the high water at any place occurs before the moon's transit, the interval between the successive high waters is less than the average of 12 hours 25 minutes and the tide is said to prime. For positions of moon between the first quarter and the full moon, the high water at any place occurs after the moon transits, the interval between successive high waters is more than the average, and tide is said to lag. Similarly, between full moon and 3rd quarter position, the tide primes while between the 3rd quarter and full moon position, the tide lags. At first quarter, full moon and third quarter position of moon, normal tide occurs.

Due to the several assumptions made in the equilibrium theory, and due to several other factors affecting the magnitude and period of tides, close agreement between the results of the theory, and the actual field observations is not available. Due to obstruction of land masses, tide may be heaped up at some places. Due to inertia and viscosity of sea water, equilibrium figure is not achieved instantaneously. Hence prediction of the tides at a place must be based largely on observations.

4.3. Astronomical Surveying

CELESTIAL SPHERE

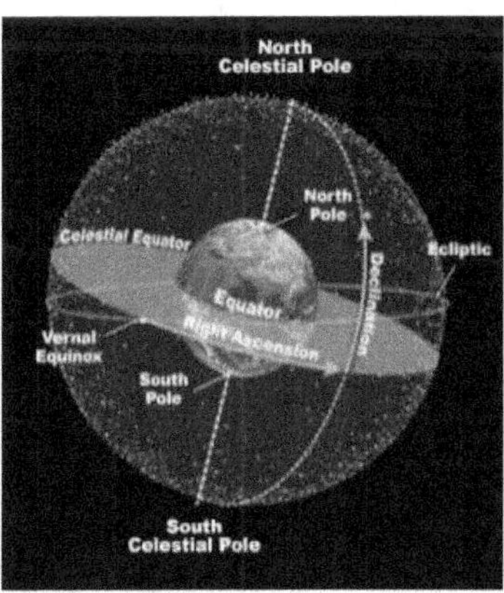

Figure 4.3.1 Celestial Sphere

4.3.1. Celestial Sphere

The millions of stars that we see in the sky on a clear cloudless night are all at varying distances from us. Since we are concerned with their relative distance rather than their actual distance from the observer. It is exceedingly convenient to picture the stars as distributed over the surface of an imaginary spherical sky having its center at the position of the observer. This imaginary sphere on which the star appear to lie or to be studded is known as the celestial sphere. The radius of the celestial sphere may be of any value - from a few thousand metres to a few thousand kilometres. Since the stars are very distant from us, the center of the earth may be taken as the center of the celestial sphere.

4.3.2. Zenith, Nadir and Celestial Horizon

The Zenith (Z) is the point on the upper portion of the celestial sphere marked by plumb line above the observer. It is thus the point on the celestial sphere immediately above the observer's station.

The Nadir (Z') is the point on the lower portion of the celestial sphere marked by the plum line below the observer. It is thus the point on the celestial sphere vertically below the observer's station. Celestial Horizon. (True or Rational horizon or geocentric horizon): It is the great circle traced upon the celestial sphere by that plane which is perpendicular to the Zenith-Nadir line, and which passes through the center of the earth. (Great circle is a section of a sphere when the cutting plane passes through the center of the sphere).

4.3.3. Terrestrial Poles and Equator, Celestial Poles and Equator

The terrestrial poles are the two points in which the earth's axis of rotation meets the earth's sphere. The terrestrial equator is the great circle of the earth, the plane of which is at right angles to the axis of rotation. The two poles are equidistant from it.

If the earth's axis of rotation is produced indefinitely, it will meet the celestial sphere in two points called the north and south celestial poles (P and P'). The celestial equator is the great circle of the celestial sphere in which it is intersected by the plane of terrestrial equator.

1) Co-Altitude or Zenith Distance (Z) and Azimuth (A)

It is the angular distance of heavenly body from the zenith. It is the complement or the altitude, i.e z = (90°).

The azimuth of a heavenly body is the angle between the observer's meridian and the vertical circle passing through the body.

CO – ORDINATE SYSTEMS

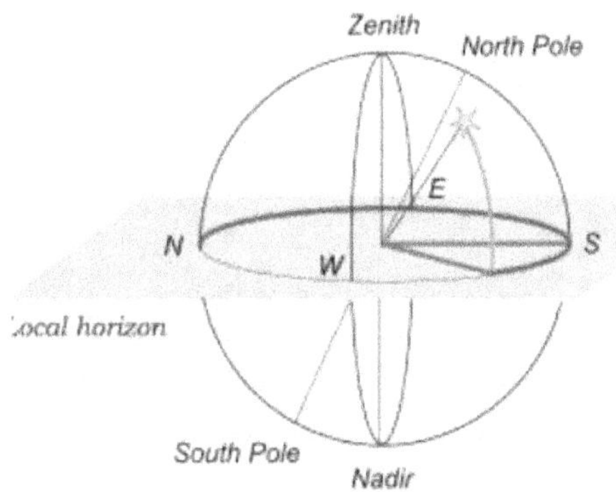

Figure 4.3.2: Coordinate Systems

Determine the hour angle and declination of a star from the following data: (i) Altitude of the star = 22°36'

(ii) Azimuth of the star = 42°W

(iii) Latitude of the place of observation = 40°N.

Solution

Since the azimuth of the star is 42°W, the star is in the western hemi-sphere. In the astronomical DPZM, we have PZ = co-latitude = 90° - 40° = 50°;

ZM = co-altitude = 90° - 22° 36° = 67 24'; angle A = 42°

Knowing the two sides and the included angle, the third side can be calculated from the cosine formula

Thus, cos PM = cos PZ. cos ZM + sin PZ. Sin ZM. cos A

= cos 50°. cos 67° 24' + sin 50°. sin 67° 24'. cos 42°

= 0.24702 + 0.52556 = 0.77258

\ PM = 39°25'

\Declination of the star = d= 90° - PM = 90° - 39°25' = **50°35' N.**

Similarly, knowing all the three sides, the hour angle H can be calculated from Eq. 1.2.

$$\cos H = \frac{\cos ZM - \cos PZ.\cos PM}{\sin PZ.\sin PM} = \frac{\cos 67°24' - \cos 50°.\cos 39°25'}{\sin 50°.\sin 39°25'}$$

$$= \quad = \frac{0.38430 - 0.49659}{0.48640} = -0.23086$$

\ cos (180° - H) = 0.23086

\ 180° - H = 76° 39'

H = 103°21'.

The tangents A, B & AC intersect at point B at a chainage 150.5 m. Calculate all the necessary data for setting out a circular curve of radius 100 m and deflection angle 30° by the method of offsets from chord. The normal chord is 30m.

Given Data

Radius R = 100 m

Angle φ = 30°

Chainage = 150.50 m

Solution

Tangent length BT1   **= R tan φ/2**

                                   = 100 (tan 30°/2)

                                     = 26.79 m

Chainage at tangent point = chainage – BT1

                             = 150.50 -26.79

                             = 123.71 m

Length of the curve = πR φ/180°

                        = π x 30x 100/180°

                        = 52.36 m

Chain age of tangent point T2 = (123.71 + 52.36)

                           = 176.07 m

Length of long chord L = 2R sin φ/2

                        = 2(100)sin 30°/2

                        = 51.76 m

Half of long chord = L/2 = 51.76/2

                        = 25.88 mm

MODEL SMALL QUESTION AND ANSWERS

ADVANCED TOPICS IN SURVEYING

PART A

1. What are the equipments used for sounding. (April/May 2018) (Nov/Dec 2012)

- Sounding rods or poles
- Sounding cables or lead lines
- Eco-sounder or Fathometer

2. Define points of tangency. (Nov/Dec 2017)

In geometry, the tangent line (or simply tangent) to a plane curve at a given point is the straight line that "just touches" the curve at that point. Leibniz defined it as the line through a pair of infinitely close points on the curve.

3. What do you understand by echo-sounding? (Nov/Dec 2017) (May/June 2014)

- This is also called as fathometer which is used for measuring depth of large rivers and of seas with depth more than 10m. By this instrument the depth of water is obtained by sending a sound impulse from the surface of water towards the bottom of the river or seabed.

4. What are the main functions of transition curve? (April/May 2017)

Primary functions of a transition curves (or easement curves) are: To accomplish gradual transition from the straight to circular curve, so that curvature changes from zero to a finite value. To changing curvature in compound and reverse curve cases, so that gradual change of curvature introduced from curve to curve.

5. Define hydro graphic surveying. (April/May 2017)

- Hydro graphic survey is the science of measurement and description of features which affect maritime navigation, marine construction, dredging, offshore oil exploration/ offshore oil drilling and related activities. Strong emphasis is placed on soundings, shorelines, tides, currents, seabed and submerged obstructions that relate to the previously mentioned activities.

6. What is the three point problem in hydro graphic surveying? (Nov/Dec 2016)

- In surveying, a method used to orient underground workings via three plumb lines suspended in a vertical shaft. The problem of determining dip and strike of a plane from elevations determined at three known points not in a straight line.

7. **What is the meant by scale of photograph? (Nov/Dec 2016)**

- Scale is an important describing factor of vertical aerial photography. It is important to know the scale of the image under examination, as this can affect how you perceive or interpret what appears in the image. Scale also allows features in the image to be measured.

8. **Define MSL. (April/May 2015)**

- Mean sea level is the sea level obtained considering the average heights of all the tides being measured at hourly intervals over some stated period covering the entire complete tides. For all major works the datum selected is the mean sea level (MSL).

9. **What is meant by three point problem in hydro graphic surveying? (Nov/Dec 2014)**

- If a sounding is located by two angles from the boat by observations to three known points on the shore, the plotting can be done adopting three-point problem. The three point problem may be solved by mechanical, graphical or analytical methods.

10. **Define Azimuth. (May/June 2013)**

- Azimuth of a heavenly body is the angle between the observer's meridian and the vertical circle passing through the body.

11. **Distinguish between the Zenith and the Nadir. (April/May 2011)**

- Zenith is the point on the upper portion of celestial sphere immediately above the overhead of an observer.
- Nadir is the intersection of a 23ertical line through the observer's station to the lower portion of the celestial sphere.

12. **Distinguish between terrestrial photogrammetry and aerial photogrammetry. (April/May 2011)**

- Photographs are taken either from ground station or from the air. Photographs taken from a fixed position on or near the ground and the branch deals on such aspects is called terrestrial photogrammetry.
- Aerial photogrammetry is that type of photo grammetry wherein the photographs are taken by cameras mounted on an aircraft flying over the area.

13. **Distinguish between a compound curve and a reverse curve.**

- A compound curve consists of a combination of two circular of different radius with a common junction. The different of change of curvature is on the same side.

- A reverse curve is basically a compound curve with a common tangent at the junction. It consists of two circular arcs turning in opposite directions with the common at the junction.

14. What is super elevation?

In order to counter balance the centrifugal force the outer edge of the road is raised which is known as the super-elevation or cant or banking. This traverse slope is provided throughout the length of the horizontal curve. The super elevation is expressed as the ratio of the height of the outer edge with respect to the horizontal width.

15. What is a transition curve?

Transition curve is also called as an basement curve which is an arc introduced between a straight and a circular curve or between two arcs of a compound curve. The radius of a transition curve varies from infinity to a fixed value.

16. Write down the requirements of an ideal transition curve.

- The transition should be tangential to the straight.
- The curvature of the transition curve should be zero at the origin of the straight.
- The exact amount of super elevation should be attained at the junction with the circular curve.
- The curvature of the transition curve should increase at the same rate as that of the super elevation

17. What is a summit curve and how it occurs.

A vertical curve with convexity upwards is called a summit curve.

- An ascending of gradient meets another ascending gradient.
- An ascending gradient intersects a descending gradient.
- A descending gradient meets another descending gradient.
- An ascending gradient meets a horizontal.

18. What is a route survey? What is its purpose?

Route surveying is applied to the surveys required to establish the horizontal and vertical alignments for transport facilities. The transport facilities may be highways, railway, aqua ducts, canals, water pipeline oil and gas lines, cableways, waterways, power, telephone and waste water disposal.

19. What is Reconnaissance survey?

Preliminary inspection of an area to be surveys is called reconnaissance or a reconnoiter. During the survey a proper planning should be done such that the work will be better and effectively executed.

20. What are lunar and solar tides?

- The periodical variations in natural water level are called as tides. The resultant force between the earth and moon causes lunar tides. Lunar tides may be super lunar tide or inferior lunar tides depending on the moon`s transit.

- The phenomenon of production of solar tides is due to force of attraction between earth and sun which is similar to the lunar tides. Thus there will be superior solar tide or inferior solar tide.

21. What is meant by sounding?

The measurements of depths below the water surface are called soundings. This is synonymous to the depth measurement in land with reference to a datum. The aim in making soundings is to determine the configuration of the subaqueous source.

22. What are three point problems in hydro graphic surveying?

If a sounding is located by two angles from the boat by observations to three known points on the shore, the plotting can be done adopting three-point problem. The three point problem may be solved by mechanical, graphical or analytical methods.

23. What is a great circle?

If the earth is considered as a sphere any plane passing through its centre traces out upon the surface a circle called great circle. For example equator is a great circle.

24. What is meant by celestial sphere?

For an observer upon the earth the fixed starts seem to be studded over the surface of a vast sphere, known as the celestial sphere at the centre of which the earth is approximately situated.

Because of the real rotation of the earth about its polar axis every twenty four hours, the celestial sphere appears to rotate about the same axis during that time. The centre sphere the earth may be taken as the centre of the celestial sphere.

25. Name the properties of spherical triangle.

- Any angle is less than two right angles or π.

- Sum of the three angles is less than six right angles or 3π and greater than two right angles or π.
- Sum of any two sides is greater than the third.
- If the sum of any two angles is equal to two right angles or π, the sum of the angles opposite them is equal to two right angles.
- The smaller angle is opposite the smaller side and vice-versa.

26. What are the corrections to be applied to the observed altitude of sun?

The corrections to be applied are Instrumental corrections and Observational corrections.

Instrumental corrections are Index Error and Bubble Error corrections. Observational corrections are the following: Correction for parallax, correction for refraction, correction for dip of the horizon and correction for semi- diameter.

27. Explain the term sidereal time.

The sidereal time at any instant is the hour angle of the first point of Aries reckoned westward from 0 h to 24 h. The right ascension (R.A.) of the meridian of a place is known as the local sidereal time (L.S.T).

$$L.S.T = (R.A.\ of\ a\ star) + (westerly\ hour\ angle\ of\ a\ star)$$

28. What is meant by Mean Solar Time?

In order to circumvent the non-uniformity of apparent solar time, a fictitious body called the mean sun is introduced.

29. What is Photogrammetry?

- Photographic surveying or photo grammetry is the art of producing plans or maps from photographs. Here the photographs are taken from a suitable camera position.
- Photographic surveying, in principles, is very similar to that of plane table surveying, with the difference that most of the work, which the latter instrument is executed in the field, is here done in the office.

30. List the uses of photogrammetry.

- Construction of plan metric and topographic maps.
- Mountains and hilly areas with less number of trees can be very satisfactorily surveyed.
- Aerial surveying is most suitable for reconnaissance.
- Acquisition of military intelligence.
- Interpretation of geology and soil details.

31. What is meant by scale of photograph?

Scale of a photograph is obtained from the ratio of the distance of any two points on the photograph and the distance between the corresponding points on the ground. The two points chosen for scaling should lie nearly equidistant on either side of the principal point.

32. Differentiate between Tilted photograph and oblique photograph.

- A tilted photograph is an aerial photograph made with the camera axis unintentionally. The tilt from the vertical axis is usually less than 3°.
- An oblique photograph is the one made in an aerial photograph intentionally between the horizontal and the vertical.

MODEL DETAIL QUESTIONS

1. Explain the various sounding methods. **(April/May 2017).**
2. What are the applications of remote sensing? **(April/May 2017) (Nov/Dec 2015)**
3. Calculate the sun's azimuth and hour angle at sunset at a place in latitude 52^0N, when its declination is,

   (i) 20^0N

   (ii) 14^0S(May/June 2014)

4. (a) Determine the hour angle and declination of a star from the following data.

      Altitude of the star = $21^030'$

      Azimuth of the star =140^0E

      Latitude of the observer = 48^0N

   (b) What are the applications of photogrammetry?

5. (a) How to measure angles with the sextant?

   (b) The following observations were made on three shore stations A,B and C from a sounding boat at P. Stations B and P are on the same side of AC. If angle ∟APB= $30^023'$, angle ∟BPC= $40^036'$and angle ∟ABC= $125^012'$. The distance AB= 4220m, BC= 5050m. Determine AP, BP and CP.

6. Explain clearly how would you determine the levels at river bed points and fix the position of sounding.

   (i)   By use of sextant in a boat.

   (ii)   By use of a theodolite onshore.

7. Write short notes on: **(Nov/Dec 2016)**

      1. Aerial photograph

      2. Stereoscopy

      3. EDM

8. What is meant by three point problem in hydrographic surveying? **(Nov/Dec 2014)**.

9. Define hydrographic surveying. Explain the various hydrographic surveying methods.

10. What are the equipments used for sounding? Explain with detail answer. **(Nov/Dec 2012)**.

5. MODERN SURVEYING

Total Station: Advantages - Fundamental quantities measured - Parts and accessories - working principle - On board calculations - Field procedure - Errors and Good practices in using Total Station GPS Surveying: Different segments - space, control and user segments - satellite configuration - signal structure - Orbit determination and representation - Anti Spoofing and Selective Availability - Task of control segment - Hand Held and Geodetic receivers - data processing - Traversing and triangulation.

5.1. Total Station: Basic Principle

Although taping and theodolites are used regularly on site - total stations are also used extensively in surveying, civil engineering and construction because they can measure both distances and angles. A typical total station is shown in the figure below.

Fig. 5.1: Total Station

Because the instrument combines both angle and distance measurement in the same unit, it is known as an integrated total station which can measure horizontal and vertical angles as well as slope distances.

Using the vertical angle, the total station can calculate the horizontal and vertical distance components of the measured slope distance. As well as basic functions, total stations are able to perform a number of different survey tasks and associated calculations and can store large amounts of data.

As with the electronic theodolite, all the functions of a total station are controlled by its microprocessor, which is accessed thought a keyboard and display.

To use the total station, it is set over one end of the line to be measured and some reflector is positioned at the other end such that the line of sight between the instrument and the reflector is unobstructed (as seen in the figure below).

- The reflector is a prism attached to a detail pole.
- The telescope is aligned and pointed at the prism.
- The measuring sequence is initiated and a signal is sent to the reflector and a part of this signal is returned to the total station.
- This signal is then analyzed to calculate the slope distance together with the horizontal and vertical angles.
- Total stations can also be used without reflectors and the telescope is pointed at the point that needs to be measured.
- Some instruments have motorized drivers and can be use automatic target recognition to search and lock into a prism - this is a fully automated process and does not require an operator.
- Some total stations can be controlled from the detail pole, enabling surveys to be conducted by one person.

Fig. 5.2: Measuring with a Total Station

Most total stations have a distance measuring range of up to a few kilometres, when using a prism, and a range of at least 100m in reflector less mode and an accuracy of 2-3mm at short ranges, which will decrease to about 4-5mm at 1km.

Although angles and distances can be measured and used separately, the most common applications for total stations occur when these are combined to define position in control surveys.

As well as the total station, site surveying is increasingly being carried out using GPS equipment. Some predictions have been made that this trend will continue, and in the long run GPS methods may replace other methods.

Although the use of GPS is increasing, total stations are one of the predominant instruments used on site for surveying and will be for some time.

Developments in both technologies will find a point where devices can be made that complement both methods.

5.2. Classification of Total Stations

5.2.1. Electro - Optical System

1. Distance Measurement

When a distance is measured with a total station, am electromagnetic wave or pulse is used for the measurement - this is propagated through the atmosphere from the instrument to reflector or target and back during the measurement.

 Distances are measured using two methods: the phase shift method, and the pulsed laser method.

This technique uses continuous electromagnetic waves for distance measurement although these are complex in nature, electromagnetic waves can be represented in their simplest from as periodic waves.

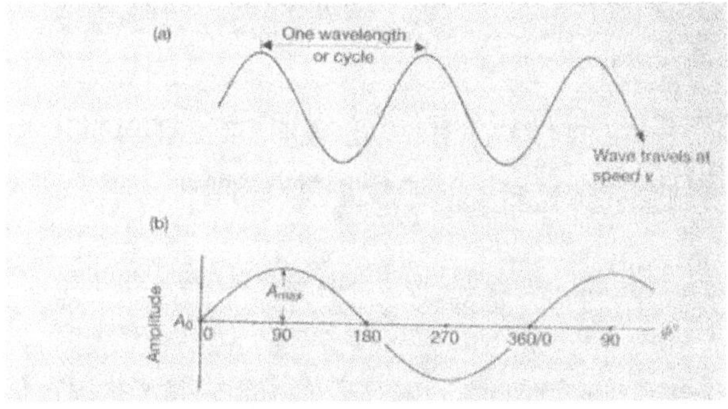

Figure 5.3: Sinusoidal Wave Motion

The wave completes a cycle when moving between identical points on the wave and the number of times in one second the wave completes the cycle is called the frequency of the wave. The speed of the wave is then used to estimate the distance.

5.2.2. Laser Distance Measurement

In many total stations, distances are obtained by measuring the time taken for a pulse of laser radiation to travel from the instrument to a prism (or target) and back. As in the phase shift method, the pulses are derived an infrared or visible laser diode and they are transmitted through the telescope towards the remote end of the distance being measured, where they are reflected and returned to the instrument.

Since the velocity v of the pulses can be accurately determined, the distance D can be obtained using $2D = vt$, where t is the time taken for a single pulse to travel from instrument - target - instrument.

This is also known as the timed-pulse or time-of-flight measurement technique.

The *transit time t* is measured using electronic signal processing techniques. Although only a single pulse is necessary to obtain a distance, the accuracy obtained would be poor. To improve this, a large number of pulses (typically 20,000 every second) are analysed during each measurement to give a more accurate distance.

The pulse laser method is a much simpler approach to distance measurement than the phase shift method, which was originally developed about 50 years ago.

5.2.3. Slope and Horizontal Distances

Both the phase shift and pulsed laser methods will measure a slope distance L from the total station along the line of sight to a reflector or target. For most surveys the horizontal distance D is required as well as the vertical component V of the slope distance.

Horizontal distance $D = L \cos? = L \sin z$

Vertical distance $= V = L \sin? = L \cos z$

Where? is the vertical angle and z is the is the zenith angle. As far as the user is concerned, these calculations are seldom done because the total station will either display D and V automatically or will dislplay L first and then D and V after pressing buttons.

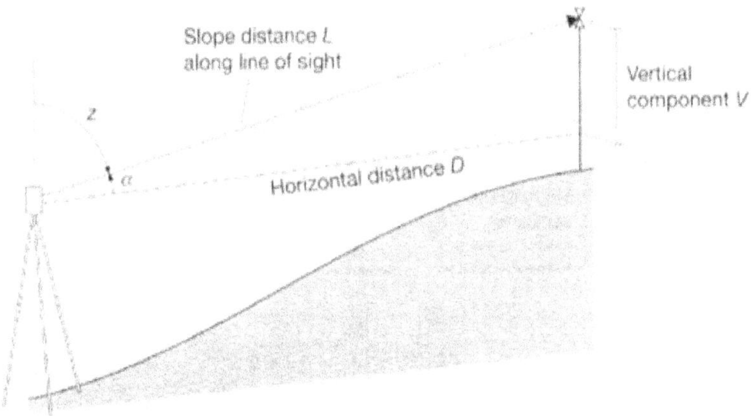

Figure 5.4: Slope and Distance Measured

How accuracy of distance measurement is specified

All total stations have a linear accuracy quoted in the form

$$\pm (a \text{ mm} + b \text{ ppm})$$

The constant a is independent of the length being measured and is made up of internal sources within the instrument that are normally beyond the control of the user. It is an estimate of the individual errors caused by such phenomena as unwanted phase shifts in electronic components, errors in phase and transit time measurements.

The systematic error b is proportional to the distance being measured, where 1 ppm (part per million) is equivalent to an additional error of 1mm for every kilometre measured.

Typical specifications for a total station vary from $\pm$ (2mm + 2ppm) to $\pm$ (5mm + 5 pmm).

For example: $\pm$ (2mm + 2ppm), at 100m the error in distance measurement will be $\pm$2mm but at 1.5km, the error will be $\pm$ (2mm + [2mm/km * 1.5km]) = $\pm$5m m.

5.2.4. Reflectors used in Distance Measurement

Since the waves or pulses transmitted by a total station are either visible or infrared, a plane mirror could be used to reflect them. This would require a very accurate alignment of the mirror, because the transmitted wave or pulses have a narrow spread.

To get around this problem special mirror prisms are used as shown below.

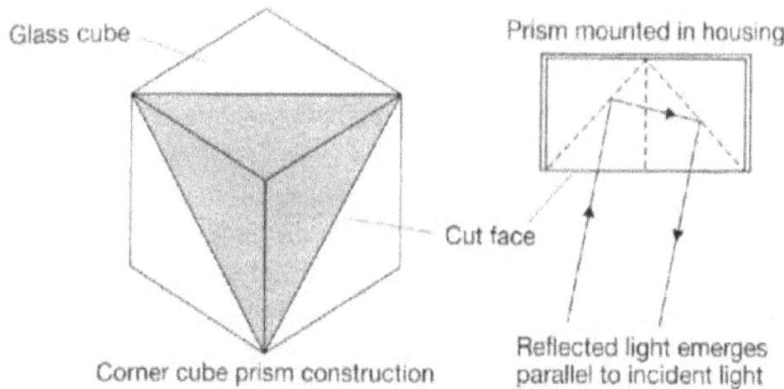

Figure 5.5: Reflector Used in Total Station

5.3. Features of Total Stations

Total stations are capable of measuring angles and distances simultaneously and combine an electronic theodolite with a distance measuring system and a microprocessor.

5.3.1. Angle Measurement

All the components of the electronic theodolite described in the previous lectures are found total stations.

The axis configuration is identical and comprises the vertical axis, the tilting axis and line of sight (or collimation). The other components include the tribatch with levelling foot screws, the keyboard with display and the telescope which is mounted on the standards and which rotates around the tilting axis.

Levelling is carried out in the same way as for a theodolite by adjusting to centralise a plate level or electronic bubble. The telescope can be transited and used in the face left (or face I) and face right (or face II) positions. Horizontal rotation of the total station about the vertical axis is controlled by a horizontal clamp and tangent screw and rotation of the telescope about the tilting axis.

The total station is used to measure angles in the same way as the electronic theodolite.

5.3.2. Distance Measurement

All total stations will measure a slope distance which the onboard computer uses, together with the zenith angle recorded by the line of sight to calculate the horizontal distance.

For distances taken to a prism or reflecting foil, the most accurate is precise measurement.

For phase shift system, a typical specification for this is a measurement time of about 1-2s, an accuracy of (2mm + 2ppm) and a range of 3-5km to a single prism.

Although all manufacturers quote ranges of several kilometres to a single prism.

For those construction projects where long distances are required to be measured, GPS methods are used in preference to total stations. There is no standard difference at which the change from one to the other occurs, as this will depend on a number of factors, including the accuracy required and the site topography.

Rapid measurement reduces the measurement time to a prism to between 0.5 and 1's for both phase shift and pulsed systems, but the accuracy for both may degrade slightly.

Tracking measurements are taken extensively when setting out or for machine control, since readings are updated very quickly and vary in response to movements of the prism which is usually pole-mounted. In this mode, the distance measurement is repeated automatically at intervals of less than 0.5s.

For reflector less measurements taken with a phase shift system, the range that can be obtained is about 100m, with a similar accuracy to that obtained when using a prism or foil.

Keyboard, Display and Software Applications in Survey Total Station.

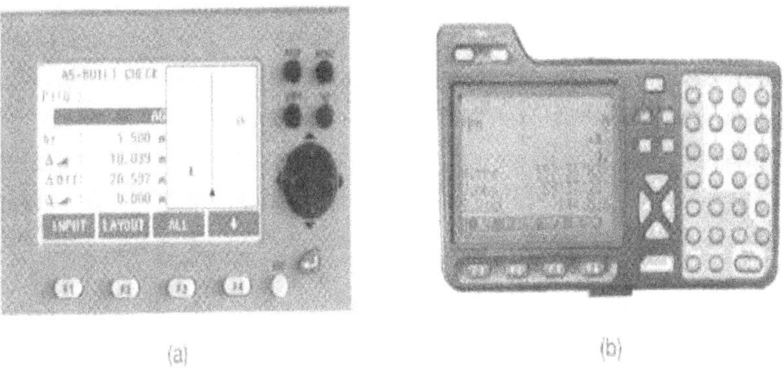

Figure 5.6: Key Board and Display

A total station is activated through its control panel, which consists of a keyboard and multiple line LCD. A number of instruments have two control panels, one on each face, which makes them easier to use.

5.3.3. Keyboard and Display

A total station is activated through its control panel, which consists of a keyboard and multiple line LCD. A number of instruments have two control panels, one on each face, which makes them easier to use.

In addition to controlling the total station, the keyboard is often used to code data generated by the instrument - this code will be used to identify the object being measured.

On some total stations it is possible to detach the keyboard and interchange them with other total stations and with GPS receivers. This is called integrated surveying.

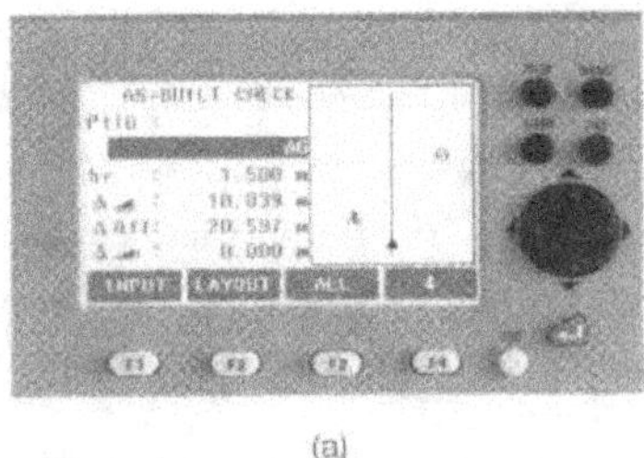
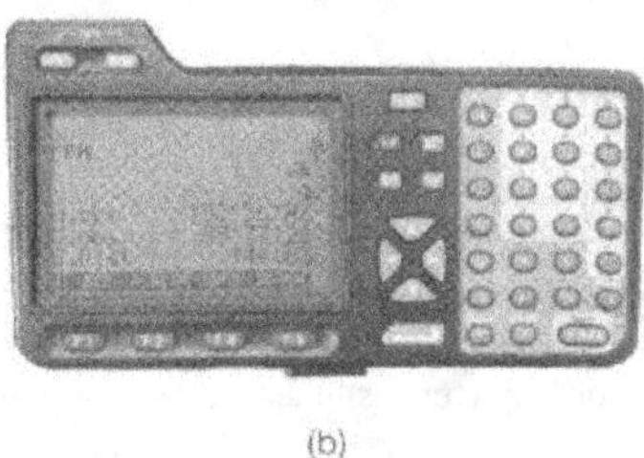

(a)   (b)

Figure 5.6.1: Key Board and Display

5.3.4. Software Applications

The microprocessor built into the total station is a small computer and its main function is controlling the measurement of angles and distances. The LCD screen guides the operator while taking these measurements.

The built in computer can be used for the operator to carry out calibration checks on the instrument.

The software applications available on many total stations include the following:

Slope corrections and reduced levels.

Horizontal circle orientation.

Coordinate measurement.

Traverse measurements.

Resection (or free stationing).

Missing line measurement.

Remote elevation measurement areas.

Setting out.

Although all manufacturers quote ranges of several kilometres to a single prism.

For those construction projects where long distances are required to be measured, GPS methods are used in preference to total stations. There is no standard difference at which the change from one to the other occurs, as this will depend on a number of factors, including the accuracy required and the site topography.

Rapid measurement reduces the measurement time to a prism to between 0.5 and 1's for both phase shift and pulsed systems, but the accuracy for both may degrade slightly.

Tracking measurements are taken extensively when setting out or for machine control, since readings are updated very quickly and vary in response to movements of the prism which is usually pole-mounted. In this mode, the distance measurement is repeated automatically at intervals of less than 0.5s.

For reflector less measurements taken with a phase shift system, the range that can be obtained is about 100m, with a similar accuracy to that obtained when using a prism or foil.

Sources of Error for Total Stations

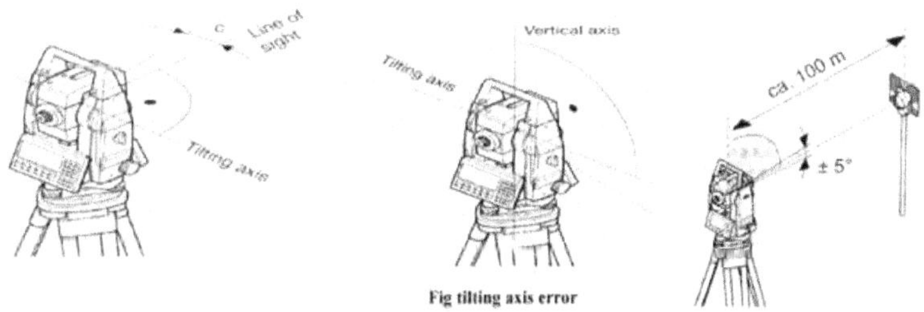

Figure 5.7: Line of Sight Error

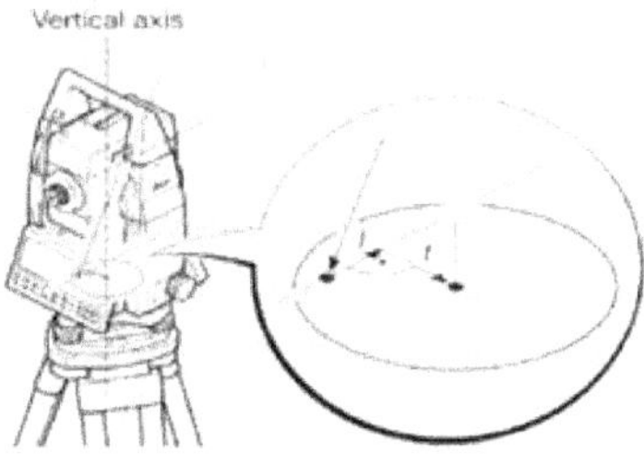

Figure 5.8: Compensator Index Error

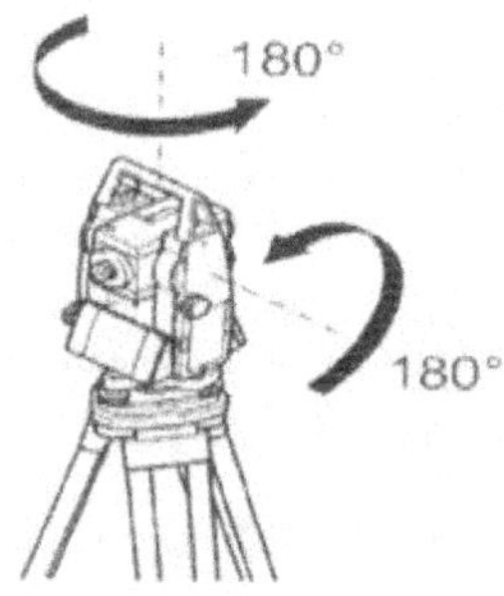

Figure 5.9: Compensator Index Error

1. Calibration of Total Stations, 2. Horizontal Collimation (Or Line Of Sight Error), 3. Tilting Axis Error, 4. Compensator Index Error.

5.4. Sources of Error for Total Stations

1) Calibration of Total Stations

To maintain the high level of accuracy offered by modern total stations, there is now much more emphasis on monitoring instrumental errors, and with this in mind, some construction sites require all instruments to be checked on a regular basis using procedures outlined in the quality manuals.

Some instrumental errors are eliminated by observing on two faces of the total station and averaging, but because one face measurements are the preferred method on site, it is important to determine the magnitude of instrumental errors and correct for them.

For total stations, instrumental errors are measured and corrected using electronic calibration procedures that are carried out at any time and can be applied to the instrument on site. These are preferred to the mechanical adjustments that used to be done in labs by technician.

Since calibration parameters can change because of mechanical shock, temperature changes and rough handling of what is a high-precision instrument, an electronic calibration should be carried our on a total station as follows:

Before using the instrument for the first time.

After long storage periods.

After rough or long transportation.

After long periods of work.

Following big changes in temperature.

Regularly for precision surveys.

Before each calibration, it is essential to allow the total station enough to reach the ambient temperature.

2) Horizontal Collimation (Or Line of Sight Error)

This axial error is caused when the line of sight is not perpendicular to the tilting axis. It affects all horizontal circle readings and increases with steep sightings, but this is eliminated by observing on two faces. For single face measurements, an on-board calibration function is used to determine c, the deviation between the actual line of sight and a line perpendicular to the tilting axis. A correction is then applied automatically for this to all horizontal circle readings.

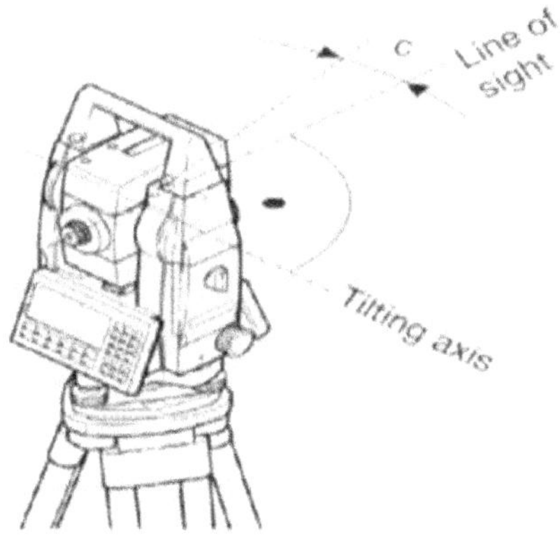

Figure 5.10: Line of Sight Error

3) Tilting Axis Error

This axial errors occur when the titling axis of the total station is not perpendicular to its vertical axis. This has no effect on sightings taken when the telescope is horizontal, but introduces errors into horizontal circle readings when the telescope is tilted, especially for steep sightings. As with horizontal collimation error, this error is eliminated by two face measurements, or the tilting axis error a is measured in a calibration procedure and a correction applied for this to all horizontal circle readings - as before if a is too big, the instrument should be returned to the manufacture.

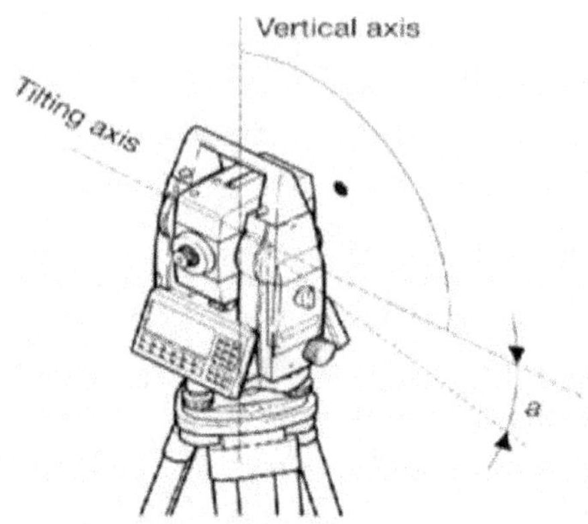

Figure 5.11: Tilting Axis Error

4) Compensator Index Error

Errors caused by not levelling a theodolite or total station carefully cannot be eliminated by taking face left and face right readings. If the total station is fitted with a compensator it will measure residual tilts of the instrument and will apply corrections to the horizontal and vertical angles for these.

However all compensators will have a longitudinal error l and traverse error t known as zero point errors. These are averaged using face left and face right readings but for single face readings must be determined by the calibration function of the total station.

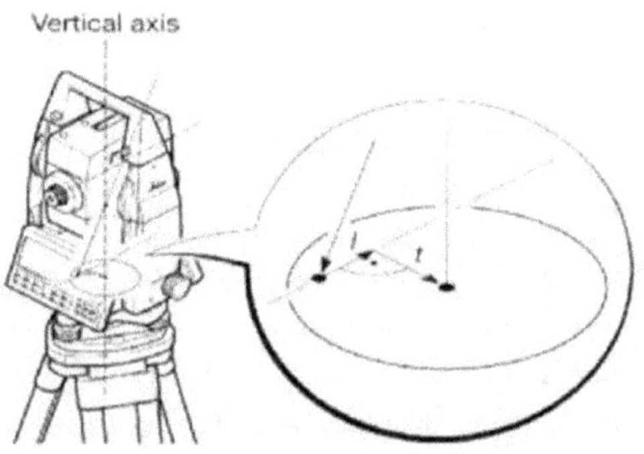

Figure 5.12: Compensator Index Error

A vertical collimation error exists on a total station if the 0° to 180° line in the vertical circle does not coincide with its vertical axis. This zero point error is present in all vertical circle readings and like the horizontal collimation error, it is eliminated by taking FL and FR readings or by determining i For all of the above total station errors (horizontal and vertical collimation, tilting axis and compensator) the total station is calibrated using an in built function. Here the function is activated and a measurement to a target is taken as shown below.

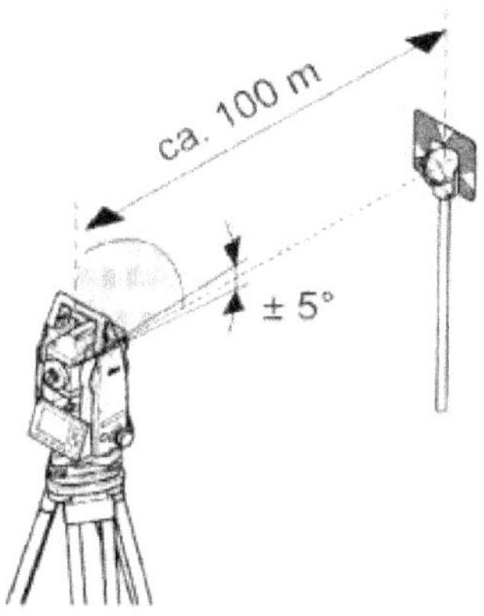

Figure 5.13

Following the first measurement the total station and the telescope are each rotated through 180° and the reading is repeated.

Any difference between the measured horizontal and vertical angles is then quantified as an instrumental error and applied to all subsequent readings automatically. The total station is thus calibrated and the procedure is the same for all of the above error type.

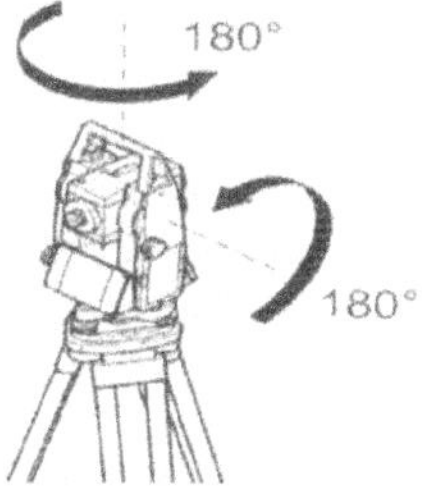

Figure 5.14: Compensator Index Error

5.5. Total Station

1. Errors

Mistakes (or) gross Errors Systematic (or) Cumulative Errors Accidental (or) Random Errors.

2. Mistakes (or) Gauss Errors

Depends upon the observer, a mistake cannot be corrected unless the observer get training. The mistakes are errors that arise from inattention, inexperience, carelessness and poor judgement of confusion in the mind of observer.

3. Systematic Errors

The systematic error is an error that under the same conditions will always be of the same size and sign. It is simply due to the error in instrument. These errors may be regarded as positive or negative according with whether they make the result too small (or) too great. This effect is cumulative.

4. Accidental Errors

The Accidental Errors are those which remain after mistakes and systematical errors have been eliminated and are caused by the combination of reasons beyond the ability of the observer to control.

5. Classification of Observer Quantity

An observer quantity may be classified as Independent Quantity Conditioned Quantity.

6. Independent Quantity

It is the one whose value is independent of the values of other quantities. It bears no relation with any other quantity and hence change in the other quantities does not affect the value of this quantity. eg. R.L of B.M.

7. Conditioned Quantity

It is the one whose value is dependent upon the values of one (or) more quantities. Its values bear a rigid relation to some other quantities. It is also called 'dependent quantities'.

8. Conditioned Equation

The conditioned equation is the equation expressing the relation existing b/w the several dependent quantities. eg. In a ABC A+B+C= 180. It is a conditioned equation.

9. Observation

An observation is a numerical value of the measured quantity and may be either direct (or) indirect.

10. Direct Observation

A direct observation is the one made directly on the quantity being determined. Eg: Measurement of base line.

11. Indirect Observation

An indirect observation is one in which the observed value is deduced from the measurement of some related quantities.

Eg: Measurement of Angle by repetition method.

12. Weight of an Observation

The weight of an observation is a number giving an indication of its precision and trust worthiness, when making a comparison between several quantities of different worth.

If a certain observation of weight 4 it means that it is 4 times as much reliable as an observation of weight 1.

When two quantities (or) observations are assumed to be equally reliable, the observed values are said to be of equal weight (or) of unit weight.

13. Weighted Observations

Observations are weighted when different weights are assigned to them. Eg: A=$30^0 40'$- wt 3. It means A is measured 3 times.

14. Observed Value of a Quantity

An observed value of a quantity is a value obtained when it is corrected for all the known errors. Observed value = Measured value ± errors (or) corrections.

15. True value of Quantity

It is the value which is obsolete free from all the errors.

16. True Error

A true error is the difference b/w the true value of the quantity and its observed value. True value = True value - observed value The most probable value of the quantity is the value which is more likely to be the true value then any other value.

17. Most Probable Errors

It is defined as the quantity which added to and subtracted from the most probable value, fixes the limit within which it is an even chance the true value of the measured quantity must lie.

18. Residual Error

It is diff b/w the most probable value of the quantity and its observed value. Residual Errors = most probable value - observed value.

19. Observation Equation

It is the relation b/w the observed quantity and its numerical value.

20. Normal Equation

It is the education which is formed by the multiplying each equation by the co-efficient of the unknown, whose normal equation is to be formed out by adding the equation thus formed.

5.6. GPS Surveying

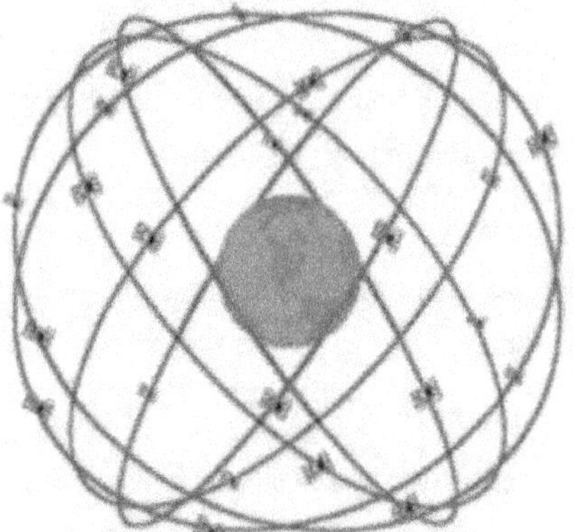

Figure 5.15: The Global Positioning System (GPS), 21 – Satellite Configuration

Traditional methods of surveying and navigation resort to tedious field and astronomical observation for deriving positional and directional information.

5.6.1. GPS Surveying

Introduction

Traditional methods of surveying and navigation resort to tedious field and astronomical observation for deriving positional and directional information. Diverse field conditions, seasonal variation and many unavoidable circumstances always bias the traditional field approach. However, due to rapid advancement in electronic systems, every aspect of human life is affected to a great deal. Field of surveying and navigation is tremendously benefited through electronic devices. Many of the critical situations in surveying/navigation are now easily and precisely solved in short time.

Astronomical observation of celestial bodies was one of the standard methods of obtaining coordinates of a position. This method is prone to visibility and weather condition and demands expert handling. Attempts have been made by USA since early 1960's to use space based artificial satellites. System TRANSIT was widely used for establishing a network of control points over large regions. Establishment of modern geocentric datum and its relation to local datum was successfully achieved through TRANSI T. Rapid improvements in higher frequently transmission and precise clock signals along with advanced stable satellite technology have been instrumental for the development of global positioning system.

The NAVSTAR GPS (Navigation System with Time and Ranging Global Positioning System) is a satellite based radio navigation system providing precise three- dimensional position, course and time information to suitably equipped user.

GPS has been under development in the USA since 1973. The US department of Defence as a worldwide navigation and positioning resource for military as well as civilian use for 24 hours and all weather conditions primarily developed it.

In its final configuration, NAVSTAR GPS consists of 21 satellites (plus 3 active spares) at an altitude of 20200 km above the earth's surface (Fig. 5.16). These satellites are so arranged in orbits to have atleast four satellites visible above the horizon anywhere on the earth, at any time of the day. GPS Satellites transmit at frequencies L1=1575.42 MHz and L2=1227.6 MHz modulated with two types of code viz. P-code and C/A code and with navigation message. Mainly two types of observable are of interest to the user. In pseudo ranging the distance between the satellite and the GPS receiver plus a small corrective term for receiver clock error is observed for positioning whereas in carrier phase techniques, the difference between the phase of the carrier signal transmitted by the satellite and the phase of the receiver oscillator at the epoch is observed to derive the precise information.

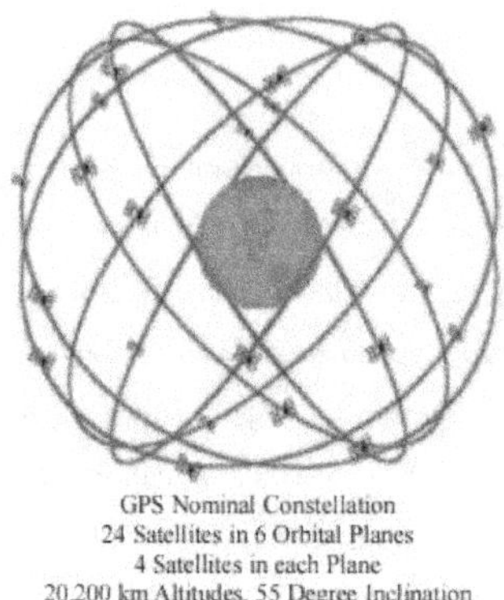

Fig. 5.16: The Global Positioning System (GPS), 21-Satellite Configuration

The GPS satellites act as reference points from which receivers on the ground detect their position. The fundamental navigation principle is based on the measurement of pseudo ranges between the user and four satellites (Fig.)

2). Ground stations precisely monitor the orbit of every satellite and by measuring the travel time of the signals transmitted from the satellite four distances between receiver and satellites will yield accurate position, direction and speed. Though three-range measurements are sufficient, the fourth observation is essential for solving clock synchronization error between receiver and satellite. Thus, the term' pseudo ranges' is derived. The secret of GPS measurement is due to the ability of measuring carrier phases to about 1/100 of a cycle equaling to 2 to 3 mm in linear distance. Moreover the high frequency L1 and L2 carrier signal can easily penetrate the ionosphere to reduce its effect. Dual frequency observations are important for large station separation and for eliminating most of the error parameters.

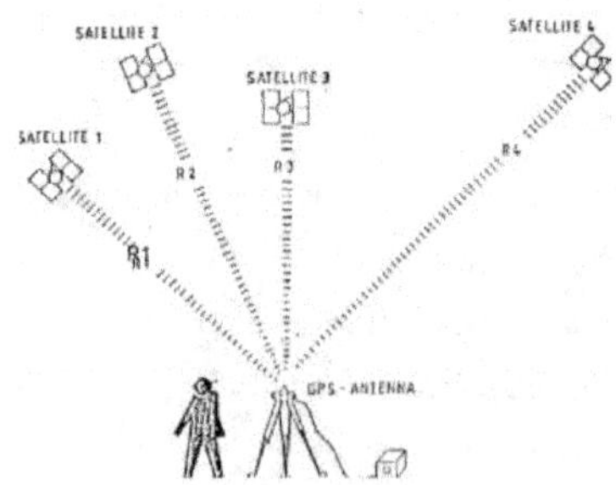

Figure 5.17: Basic Principle of Positioning with GPS

There has been significant progress in the design and miniaturization of stable clock. GPS satellite orbits are stable because of the high altitudes and no atmosphere drag. However, the impact of the sun and moon on GPS orbit though significant, can be computed completely and effect of solar radiation pressure on the orbit and tropospheric delay of the signal have been now modeled to a great extent from past experience to obtain precise information for various applications.

Comparison of main characteristics of TRANSIT and GPS re veal technological advancement in the field of space based positioning system (Table 1).

Table 1: TRANSIT vs GPS

Details	TRANSIT	GPS
Orbit Altitude	1000 Km	20,200 Km
Orbital Period	105 Min	12 Hours
Frequencies	150 MHz 400 MHz	1575 MHz 1228 MHz
Navigation data	2D : X, Y	4D : X,Y,Z, t velocity
Availability	15-20 minute per pass	Continuously
Accuracy	ñ 30-40 meters (Depending on velocity	ñ15m (Pcode/No. SA 0.1 Knots
Repeatability	—	ñ1.3 meters relative
Satellite	4-6	21-24
Geometry	Variable	Repeating
Satellite Clock	Quartz	Rubidium, Cesium

GPS has been designed to provide navigational accuracy of ±10 m to ±15 m. However, sub meter accuracy in differential mode has been achieved and it has been proved that broad varieties of problems in geodesy and geodynamics can be tackled through GPS.

Versatile use of GPS for a civilian need in following fields have been successfully practiced viz. navigation on land, sea, air, space, high precision kinematics survey on the ground, cadastral surveying, geodetic control network densification, high precision aircraft positioning, photogrammetry without ground control, monitoring deformations, hydrographic surveys, active control survey and many other similar jobs related to navigation and positioning,. The outcome of a typical GPS survey includes geocentric position accurate to 10 m and relative positions between receiver locations to centimeter level or better.

GPS Surveying

Traditional methods of surveying and navigation resort to tedious field and astronomical observation for deriving positional and directional information. Diverse field conditions, seasonal variation and many unavoidable circumstances always bias the traditional field approach. However, due to rapid advancement in electronic systems, every aspect of human life is affected to a great deal. Field of surveying and navigation is tremendously benefited through electronic devices. Many of the critical situations in surveying/navigation are now easily and precisely solved in short time.

5.6.2. Segments of GPS

For better understanding of GPS, we normally consider three major segments viz. space segment, Control segment and User segment. Space segment deals with GPS satellites systems, Control segment describes ground based time and orbit control prediction and in User segment various types of existing GPS receiver and its application is dealt.

Table 2 gives a brief account of the function and of various segments along with input and output information.

Table 2: Functions of Various Segments of GPS

Segmen	Input	Function	Output
Space	Navigation message	Generate and Transmit code and carrier	P-Code C/A Code L1,L2
Control	P-Code Observations Time	Produce GPS time predict ephemeris	Navigation message
User	Code observation Carrier phase observation	Navigation solution Surveying	Position velocity time

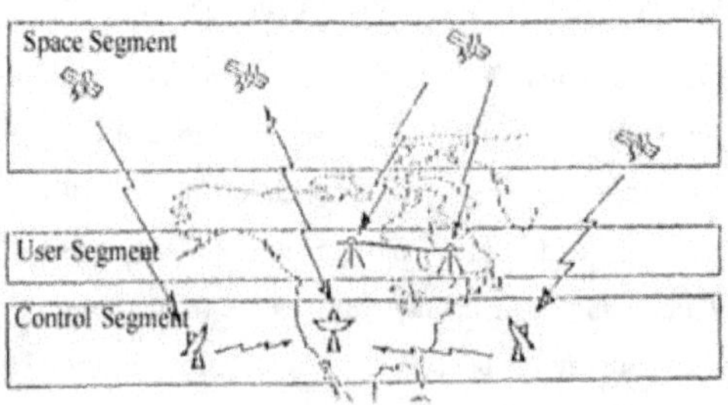

Figure 5.18: The Space, Control and User Segments of GPS

GLONASS (Global Navigation & Surveying System) a similar system to GPS is being developed by former Soviet Union and it is considered to be a valuable complementary system to GPS for future application.

GPS Surveying: Space Segment

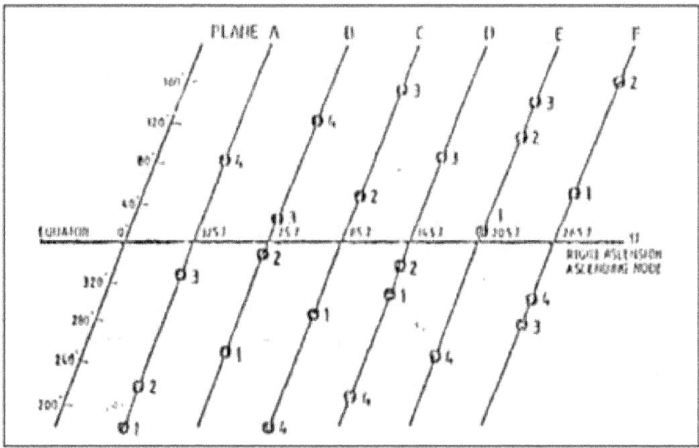

Figure 5.19: Arrangement of Satellites in Full Constellation

Space segment will consist 21 GPS satellites with an addition of 3 active spares. These satellites are placed in almost six circular orbits with an inclination of 55 degree.

GPS Surveying

5.6.3. Space Segment

Space segment will consist 21 GPS satellites with an addition of 3 active spares. These satellites are placed in almost six circular orbits with an inclination of 55 degree. Orbital height of these satellites is about 20,200 km corresponding to about 26,600 km from the semi major axis. Orbital period is exactly 12 hours of sidereal time and this provides repeated satellite configuration every day advanced by four minutes with respect to universal time.

Final arrangement of 21 satellites constellation known as 'Primary satellite constellation' is given in Fig. 5.19.1. There are six orbital planes A to F with a separation of 60 degrees at right ascension (crossing at equator). The position of a satellite within a particular orbit plane can be identified by argument of latitude or mean anomaly M for a given epoch.

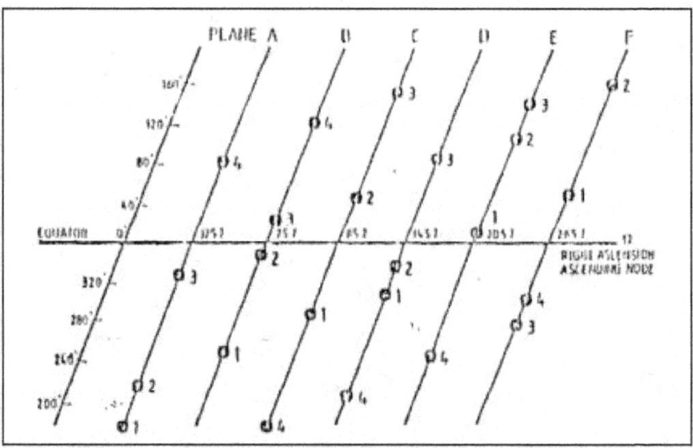

Figure 5.19.1: Arrangement of Satellites in Full Constellation

GPS satellites are broadly divided into three blocks: Block-I satellite pertains to development stage, Block II represents production satellite and Block IIR are replenishment/ spare satellite.

Under Block-I, NAVSTAR 1 to 11 satellites were launched before 1978 to 1985 in two orbital planes of 63-degree inclination. Design life of these prototype test satellites was only five years but the operational period has been exceeded in most of the cases.

The first Block-II production satellite was launched in February 1989 using channel Douglas Delta 2 booster rocket. A total of 28 Block-II satellites are planned to support 21+3 satellite configuration. Block-II satellites have a designed lifetime of 5-7 years.

To sustain the GPS facility, the development of follow-up satellites under Block-II R has started. Twenty replenishment satellites will replace the current block-II satellite as and when necessary. These GPS satellites under Block-IR have additional ability to measure distances between satellites and will also compute ephemeris on board for real time information gives a schematic view of Block-II satellite. Electrical power is generated through two solar panels covering a surface area of 7.2 square meter each. However, additional battery backup is provided to provide energy when the satellite moves into earth's shadow region. Each satellite weighs 845kg and has a propulsion system for positional stabilization and orbit maneuvers.

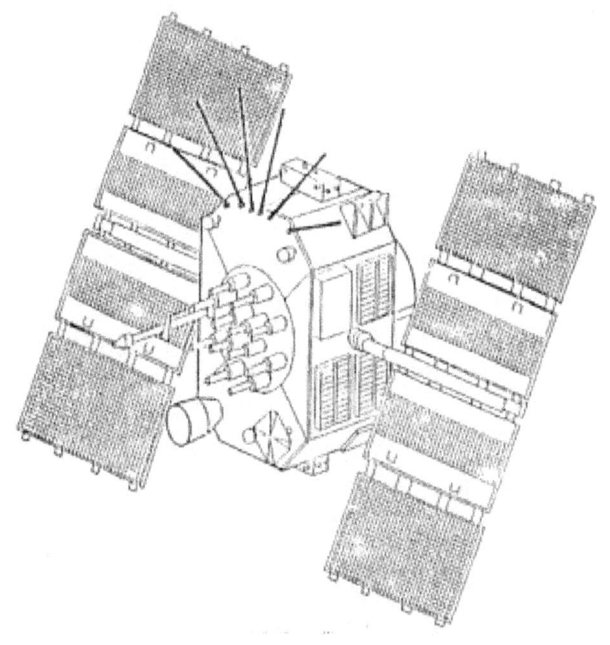

Figure 5.20: Schematic View of a Block II GPS Satellite

GPS satellites have a very high performance frequency standard with an accuracy of between $1X10^{-12}$ to $1X10^{-13}$ and are thus capable of creating precise time base. Block-I satellites were partly equipped with only quartz oscillators but Block-II satellites have two cesium frequency standards and two rubidium frequency standards. Using fundamental frequency of 10.23 MHz, two carrier frequencies are generated to transmit signal codes.

5.6.4. Observation Principle and Signal Structure

NAVSTAR GPS is a one-way ranging system i.e. signals are only transmitted by the satellite. Signal travel time between the satellite and the receiver is observed and the range distance is calculated through the knowledge of signal propagation velocity. One way ranging means that a clock reading at the transmitted antenna is compared with a clock reading at the receiver antenna. But since the two clocks are not strictly synchronized, the observed signal travel time is biased with systematic synchronization error. Biased ranges are known as pseudoranges. Simultaneous observations of four pseudoranges are necessary to determine X, Y, Z coordinates of user antenna and clock bias.

Real time positioning through GPS signals is possible by modulating carrier frequency with Pseudorandom Noise (PRN) codes. These are sequence of binary values (zeros and ones or +1 and -1) having random character but identifiable distinctly. Thus pseudoranges are derived from travel time of an identified PRN signal code. Two different codes viz. P-code and C/A code are in use. P means precision or protected and C/A means clear/acquisition or coarse acquisition.

P-code has a frequency of 10.23 MHz. This refers to a sequence of 10.23 million binary digits or chips per second. This frequency is also referred to as the chipping rate of P-code. Wavelength corresponding to one chip is 29.30m. The P-code sequence is extremely long and repeats only after 266 days. Portions of seven days each are assigned to the various satellites. As a consequence, all satellite can transmit on the same frequency and can be identified by their unique one-week segment. This technique is also called as Code Division Multiple Access (CDMA). P-code is the primary code for navigation and is available on carrier frequencies L1 and L2.

The C/A code has a length of only one millisecond; its chipping rate is 1.023 MHz with corresponding wavelength of 300 meters. C/A code is only transmitted on L1 carrier.

GPS receiver normally has a copy of the code sequence for determining the signal propagation time. This code sequence is phase-shifted in time step- by-step and correlated with the received code signal until maximum correlation is achieved. The necessary phase-shift in the two sequences of codes is a measure of the signal travel time between the satellite and the receiver antennas. This technique can be explained as code phase observation.

For precise geodetic applications, the pseudoranges should be derived from phase measurements on the carrier signals because of much higher resolution. Problems of ambiguity determination are vital for such observations.

The third type of signal transmitted from a GPS satellite is the broadcast message sent at a rather slow rate of 50 bits per second (50 bps) and repeated every 30 seconds. Chip sequence of P-code and C/A code are separately combined with the stream of message bit by binary addition i.e the same value for code and message chip gives 0 and different values result in 1.

The main features of all three signal types used in GPS observation viz carrier, code and data signals are given in Table 3.

5.6.5. GPS Satellite Signals

Table 3

GPS Satellite Signals

Atomic Clock (G, Rb) fundamental	10.23. MHz
L1 Carrier Signal	154 X 10.23 MHz
L1 Frequency	1575.42 MHz
L1 Wave length	19.05 Cm
L2 Carrier Signal	120 X 10.23 MHz
L2 Frequency	1227.60 MHz
L2 Wave Length	24.45 Cm
P-Code Frequency (Chipping Rate)	10.23 MHz (Mbps)
P-Code Wavelength	29.31 M
P-Code Period	267 days : 7
C/A-Code Frequency (Chipping Rate)	1.023 MHz (Mbps)
C/A-Code Wavelength	293.1 M
C/A-Code Cycle Length	1 Milisecond
Data Signal Frequency	50 bps
Data Signal Cycle Length	30 Seconds

The signal structure permits both the phase and the phase shift (Doppler effect) to be measured along with the direct signal propagation. The necessary bandwidth is achieved by phase modulation of the PRN code as illustrated in Fig. 5.21.

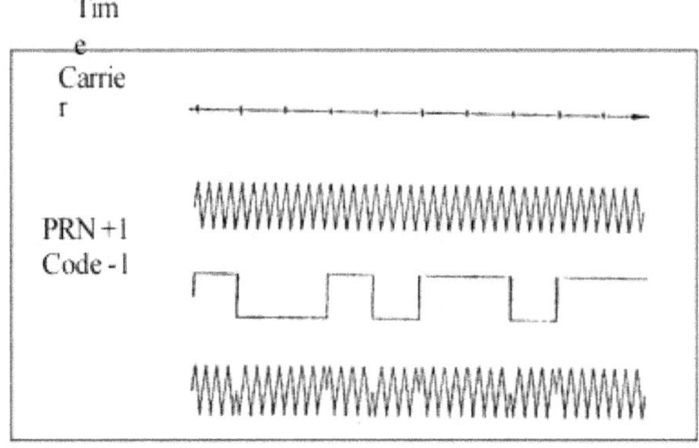

Figure 5.21: Generation of GPS Signals

Structure of the GPS Navigation Data

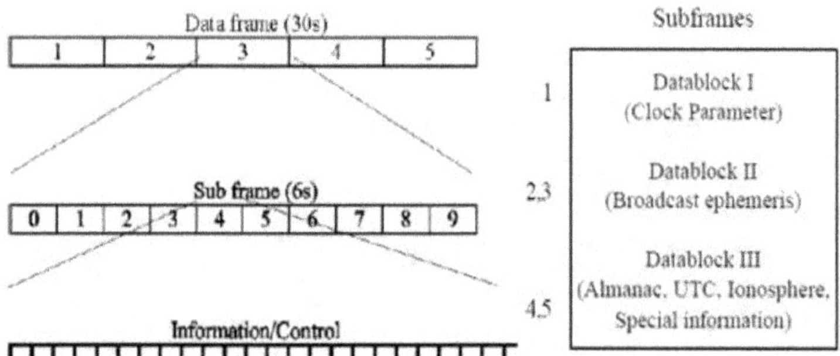

Figure 5.22: Data Block

Structure of GPS navigation data (message) is shown in Fig. The user has to decode the data signal to get access to the navigation data.

5.6.6. Structure of the GPS Navigation Data

Structure of GPS navigation data (message) is shown in Fig. 5.23. The user has to decode the data signal to get access to the navigation data. For on line navigation purposes, the internal processor within the receiver does the decoding. Most of the manufacturers of GPS receiver provide decoding software for post processing purposes. With a bit rate of 50 bps and a cycle time of 30 seconds, the total information content of a navigation data set is 1500 bits. The complete data frame is subdivided into five subframes of six-second duration comprising 300 bits of information. Each subframe contains the data words of 30 bits each. Six of these are control bits. The first two words of each subframe are the Telemetry Work (TLM) and the C/A-P-Code Hand over Work (HOW). The TLM work contains a synchronization pattern, which facilitates the access to the navigation data. Since GPS is a military navigation system of US, a limited access to the total system accuracy is made available to the civilian users. The service available to the civilians is called Standard Positioning System (SPS) while the service available to the authorized users is called the Precise Positioning Service (PPS). Under current policy the accuracy available to SPS users is 100m, 2D- RMS and for PPS users it is 10 to 20 meters in 3D. Additional limitation viz. Anti-Spoofing (AS), and Selective Availability (SA) was further imposed for civilian users. Under AS, only authorized users will have the means to get access to the P-code. By imposing SA condition, positional accuracy from Block-II satellite was randomly offset for SPS users. Since May 1, 2000 according to declaration of US President, SA is switched off for all users.

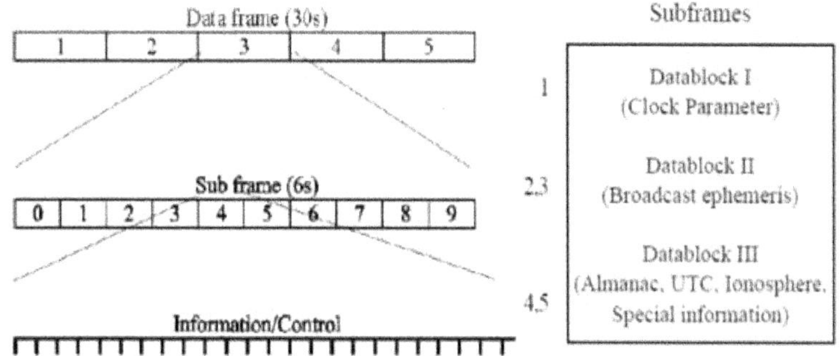

Figure 5.23: Data Block

The navigation data record is divided into three data blocks:

Data Block I appears in the first subframe and contains the clock coefficient/bias.

Data Block II appears in the second and third subframe and contains all necessary parameters for the computation of the satellite coordinates.

Data Block III appears in the fourth and fifth subframes and contains the almanac data with clock and ephemeris parameter for all available satellite of the GPSN system. This data block includes also ionospheric correction parameters and particular alphanumeric information for authorized users.

Unlike the first two blocks, the subframe four and five are not repeated every 30 seconds.

International Limitation of the System Accuracy.

The GPS system time is defined by the cesium oscillator at a selected monitor station. However, no clock parameter are derived for this station. GPS time is indicated by a week number and the number of seconds since the beginning of the current week. GPS time thus varies between 0 at the beginning of a week to 6,04,800 at the end of the week. The initial GPS epoch is January 5, 1980 at 0 hours Universal Time. Hence, GPS week starts at Midnight (UT) between Saturday and Sunday. The GPS time is a continuous time scale and is defined by the main clock at the Master Control Station (MCS). The leap seconds is UTC time scale and the drift in the MCS clock indicate that GPS time and UTC are not identical. The difference is continuously monitored by the control segment and is broadcast to the users in the navigation message. Difference of about 7 seconds was observed in July, 1992.

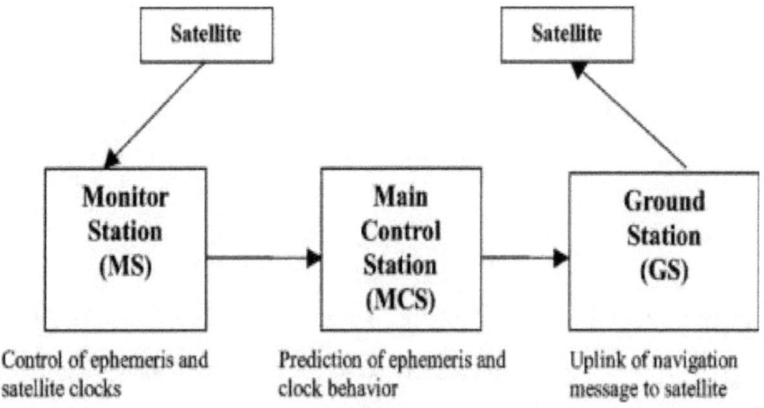

Figure 5.24: Data Flow in the Determination of the Broadcast Ephemeris

GPS satellite is identified by two different numbering schemes. Based on launch sequence, SVN (Space Vehicle Number) or NAVSTAR number is allocated. PRN (Pseudo Random Noise) or SVID (Space Vehicle Identification) number is related to orbit arrangement and the particular PRN segment allocated to the individual satellite. Usually the GPS receiver displays PRN number.

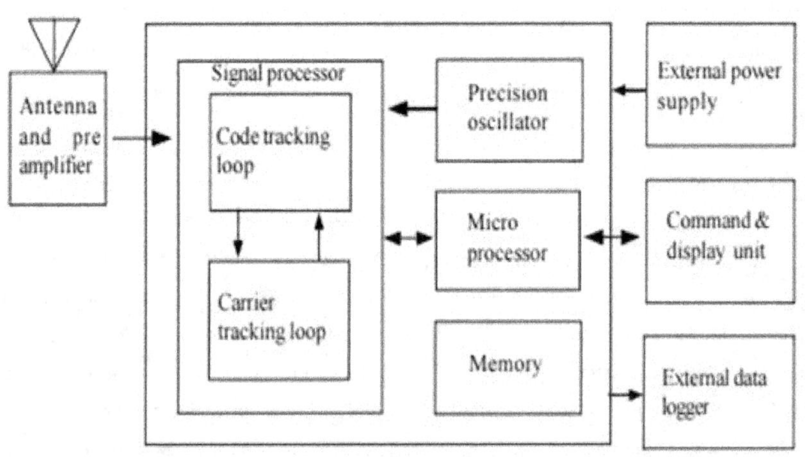

Figure 5.25: Major Components of a GPS Receiver

Basic Concept of GPS Receiver and Its Components

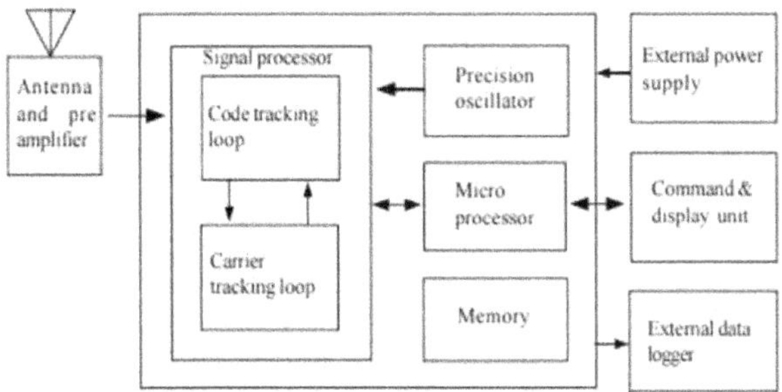

Figure 5.26: Major Components of a GPS Receiver

Antenna with pre-amplifier - RF section with signal identification and signal processing - Micro-processor for receiver control, data sampling and data processing - Precision oscillator - Power supply - User interface, command and display panel - Memory, data storage.

5.7. Basic Concept of GPS Receiver and Its Components

The main components of a GPS receiver are shown in Fig. 10. These are:

- Antenna with pre-amplifier
- RF section with signal identification and signal processing
- Micro-processor for receiver control, data sampling and data processing
- Precision oscillator
- Power supply
- User interface, command and display panel
- Memory, data storage

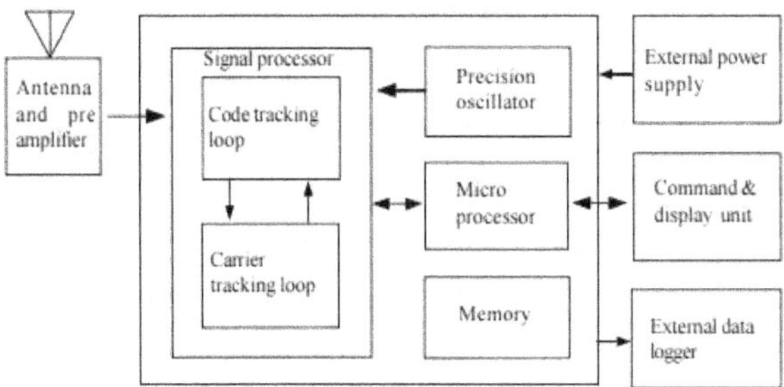

Figure 5.27: Major Components of a GPS Receiver

Antenna

Sensitive antenna of the GPS receiver detects the electromagnetic wave signal transmitted by GPS satellites and converts the wave energy to electric current] amplifies the signal strength and sends them to receiver electronics.

Several types of GPS antennas in use are mostly of following types (Fig.).

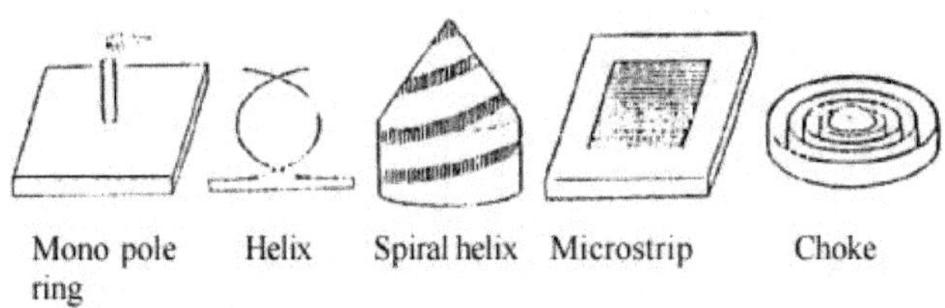

Figure 5.28: Mono Pole Helix Spiral Helix Microstrip Choke Ring

Types of GPS Antenna

- Mono pole or dipole
- Quadrifilar helix (Volute)
- Spiral helix
- Microstrip (patch)
- Choke ring

Microstrip antennas are most frequently used because of its added advantage for airborne application, materialization of GPS receiver and easy construction. However, for geodetic needs, antennas are designed to receive both carrier frequencies L1 and L2. Also they are protected against multipath by extra ground planes or by using choke rings. A choke ring consists of strips of conductor which are concentric with the vertical axis of the antenna and connected to the ground plate which in turns reduces the multipath effect.

5.7.1. RF Section with Signal Identification and Processing

The incoming GPS signals are down converted to a lower frequency in the RS section and processed within one or more channels. Receiver channel is the primary electronic unit of a GPS receiver. A receiver may have one or more channels. In the parallel channel concept each channel is continuously franking one particular satellite. A minimum of four parallel channels is required to determine position and time. Modern receivers contain upto 12 channels for each frequency.

In the sequencing channel concept the channel switches from satellite to satellite at regular interval. A single channel receiver takes atleast four times of 30 seconds to establish first position fix, though some receiver types have a dedicated channel for reading the data signal. Now days in most of the cases fast sequencing channels with a switching rate of about one-second per satellite are used.

In multiplexing channel, sequencing at a very high speed between different satellites is achieved using one or both frequencies. The switching rate is synchronous with the navigation message of 50 bps or 20 milliseconds per bit. A complete sequence with four satellites is completed by 20 millisecond or after 40 millisecond for dual frequency receivers. The navigation message is continuous, hence first fix is achieved after about 30 seconds.

Though continuous tracking parallel channels are cheap and give good overall performance, GPS receivers based on multiplexing technology will soon be available at a cheaper price due to electronic boom.

Microprocessor

`To control the operation of a GPS receiver, a microprocessor is essential for acquiring the signals, processing of the signal and the decoding of the broadcast message. Additional capabilities of computation of on-line position and velocity, conversion into a given local datum or the determination of waypoint information are also required. In future more and more user relevant software will be resident on miniaturized memory chips.

Precision Oscillator

A reference frequency in the receiver is generated by the precision oscillator. Normally, less expensive, low performance quartz oscillator is used in receivers since the precise clock information is obtained from the GPS satellites and the user clock error can be eliminated through double differencing technique when all participating receivers observe at exactly the same epoch. For navigation with two or three satellites only an external high precision oscillator is used.

Power Supply

First generation GPS receivers consumed very high power, but modern receivers are designed to consume as little energy as possible. Most receivers have an internal rechargeable. Nickel-Cadmium battery in addition to an external power input. Caution of low battery signal prompts the user to ensure adequate arrangement of power supply.

Memory Capacity

For port processing purposes all data have to be stored on internal or external memory devices. Post processing is essential for multi station techniques applicable to geodatic and surveying problems. GPS observation for pseudoranges, phase data, time and navigation message data have to be recorded. Based on sampling rate, it amount to about 1.5 Mbytes of data per hour for six satellites and 1 second data for dual frequency receivers. Modern receivers have internal memories of 5 Mbytes or more. Some receivers store the data on magnetic tape or on a floppy disk or hard-disk using external microcomputer connected through RS-232 port.

Most modern receivers have a keypad and a display for communication between the user and the receivers. The keypad is used to enter commands, external data like station number or antenna height or to select a menu operation. The display indicates computed coordinates, visible satellites, data quality indices and other suitable information. Current operation software packages are menu driven and very user friendly.

5.7.2. Classification of GPS Receivers

GPS receivers can be divided into various groups according to different criteria. In the early stages two basic technologies were used as the classification criteria viz. Code correlation receiver technology and sequencing receiver technology, which were equivalent to code dependent receivers and code free receivers. However, this kind of division is no longer justifiable since both techniques are implemented in present receivers.

Another classification of GPS receivers is based on acquisition of data types,

e.g.

- C/A code receiver
- C/A code + L1 Carrier phase
- C/A code + L1 Carrier phase + L2 Carrier phase
- C/A code + p_code + L1, L2 Carrier phase
- L1 Carrier phase (not very common)
- L1, L2 Carrier phase (rarely used)

Based on technical realization of channel, the GPS receivers can be classified as:

- Multi-channel receiver
- Sequential receiver
- Multiplexing receiver

GPS receivers are even classified on the purpose as:

- Military receiver
- Civilian receiver
- Navigation receiver
- Timing receiver
- Geodetic receiver

For geodetic application it is essential to use the carrier phase data as observable. Use of L1 and L2 frequency is also essential along with P-code.

5.7.3. Examples of GPS Receiver

GPS receiver market is developing and expanding at a very high speed. Receivers are becoming powerful, cheap and smaller in size. It is not possible to give details of every make but description of some typical receivers given may be regarded as a basis for the evaluation of future search and study of GPS receivers.

Classical Receivers

Detailed description of code dependent T1 4100 GPS Navigator and code free Macrometer V1000 is given here:

T1 4100 GPS Navigator was manufactured by Texas Instrument in 1984. It was the first GPS receiver to provide C/A and P code and L1 and L2 carrier phase observations. It is a dual frequency multiplexing receiver and suitable for geodesist, surveyor and navigators. The observables through it are:

- P-Code pseudo ranges on L1 and L2
- C/A-Code pseudo ranges on L1
- Carrier phase on L1 and L2

The data are recorded by an external tape recorder on digital cassettes or are downloaded directly to an external microprocessor. A hand held control display unit (CDU) is used for communication between observer and the receiver. For navigational purposes the built in microprocessor provides position and velocity in real time every three seconds. T1 4100 is a bulky instrument weighing about 33 kg and can be packed in two transportation cases. It consumes 90 watts energy in operating mode of 22V-32V. Generator use is recommended. The observation noise in P-Code is between 0.6 to 1 m, in C/A code it ranges between 6 to 10 m and for carrier phase it is between 2 to 3m.

T1 4100 has been widely used in numerous scientific and applied GPS projects and is still in use. The main disadvantages of the T1 4100 compared to more modern GPS equipment's are

- Bulky size of the equipment
- High power consumption
- Difficult operation procedure
- Limitation of tracking four satellites simultaneously
- High noise level in phase measurements

Sensitivity of its antenna for multipath and phase centre variation if two receivers are connected to one antenna and tracking of seven satellites simultaneously is possible. For long distances and in scientific projects, T1 4100 is still regarded useful. However, due to imposition of restriction on P-code for civilian, T1 4100 during Anti Spoofing (AS) activation can only be used as a single frequency C/A code receiver.

The MACROMETER V 1000, a code free GPS receiver was introduced in 1982 and was the first receiver for geodetic applications. Precise results obtained through it has demonstrated the potential of highly accurate GPS phase observations. It is a single frequency receiver and tracks 6 satellites on 6 parallel channels. The complete system consists of three units viz.

- Receiver and recorder with power supply
- Antenna with large ground plane
- P 1000 processor

The processor is essential for providing the almanac data because the Macrometer V 1000 cannot decode the satellite messages and process the data. At pre determined epochs the phase differences between the received carrier signal and a reference signal from receiver oscillator is measured. A typical baseline accuracy reported for upto 100 km distance is about 1 to 2 ppm (Parts per million).

Macrometer II, a dual frequency version was introduced in 1985. Though it is comparable to Macrometer V 1000, its power consumption and weight are much less. Both systems require external ephemeredes. Hence specialized operators of few companies are capable of using it and it is required to synchronize the clock of all the instruments proposed to be used for a particular observation session. To overcome above disadvantages, the dual frequency Macrometer II was further miniaturized and combined with a single frequency C/A code receiver with a brand name MINIMAC in 1986, thus becoming a code dependent receiver.

5.7.4. Examples of Present Geodetic GPS Receivers

Few of the currently available GPS receivers that are used in geodesy surveying and precise navigation are described. Nearly all models started as single frequency C/A-Code receivers with four channels. Later L2 carrier phase was added and tracking capability was increased. Now a days all leading manufacturers have gone for code-less, non- sequencing L2 technique. WILD/ LEITZ (Heerbrugg, Switzerland) and MAGNAVOX (Torrance, California) have jointly developed WM 101 geodetic receiver in 1986. It is a four channel L1 C/A code receiver. Three of the channels sequentially track upto six satellites and the fourth channel, a house keeping channels, collects the satellite message and periodically calibrates the inter channel biases. C/A-code and reconstructed L1 carrier phase data are observed once per second.

The dual frequency WM 102 was marketed in 1988 with following key features:
- L1 reception with seven C/A code channel tracking upto six satellites simultaneously.
- L2 reception of up to six satellites with one sequencing P- code channel

Modified sequencing technique for receiving L2 when P-code signals are encrypted.

The observations can be recorded on built in data cassettes or can be transferred on line to an external data logger in RS 232 or RS 422 interface. Communication between operator and receiver is established by alpha numerical control panel and display WM 101/102 has a large variety of receiver resident menu driven options and it is accompanied by comprehensive post processing software.

In 1991, WILD GPS system 200 was introduced. Its hardware comprises the Magnavox SR 299 dual frequency GPS sensor, the hand held CR 233 GPS controller and a Nicd battery. Plug in memory cards provide the recording medium. It can track 9 satellites simultaneously on L1 and L2. Reconstruction of carrier phase on L1 is through C/A code and on L2 through P-code. The receiver automatically switches to codeless L2 when P-code is encrypted. It consumes 8.5 watt through 12-volt power supply.

TRIMBLE NAVIGATION (Sunny vale, California) has been producing TRIMBLE 4000 series since 1985. The first generation receiver was a L1 C/ A code receiver with five parallel channels providing tracking of 5 satellites simultaneously. Further upgradation included increasing the number of channels upto twelve, L2 sequencing capability and P-code capability. TRIMBLE Geodatic Surveyor 4000 SSE is the most advanced model. When P-Code is available, it can perform following types of observations, viz.

- Full cycle L1 and L2 phase measurements.
- L1 and L2, P-Code measurements when AS is on and P-code is encrypted.

- Full cycle L1 and L2 phase measurement.
- Low noise L1, C/A code.
- Cross-correlated Y-Code data.

Observation noise of the carrier phase measurement when P-code is available is about ± 0-2mm and of the P-code pseudoranges as low as ± 2cm. Therefore, it is very suitable for fast ambiguity solution techniques with code/ carrier combinations.

ASHTECH (Sunnyvale, California) developed a GPS receiver with 12 parallel channels and pioneered current multi-channel technology. ASHTECH XII GPS receiver was introduced in 1988. It is capable of measuring pseudoranges, carrier phase and integrated dopler of up to 12 satellites on L1. The pseudo ranges measurement are smoothed with integrated Doppler. Postion velociy, time and navigation informations are displayed on a keyboard with a 40-characters display. L2 option adds 12 physical L2 squaring type channels.

ASHTECH XII GPS receiver is a most advanced system, easy to handle and does not require initialization procedures. Measurements of all satellites in view are carried out automatically. Data can be stored in the internal solid plate memory of 5 Mbytes capacity. The minimum sampling interval is 0.5 seconds. Like many other receivers it has following additional options viz.

- 1 ppm timing signal output.
- Photogrammetric camera input.
- Way point navigation.
- Real time differential navigation and provision of port processing and vision planning software.

In 1991, ASHTECH P-12 GPS receiver was marketed. It has 12 dedicated channels of L1, P-code and carrier and 12 dedicated channels of L2, P-code and carrier. It also has 12 L1, C/A code and carrier channels and 12 code less squaring L2 channels. Thus the receiver contains 48 channels and provides all possibilities of observations to all visible satellites. The signal to noise level for phase measurement on L2 is only slightly less than on L1 and significantly better than with code-less techniques. In cases of activated P-code encryption, the code less L2 option can be used.

TURBO ROGUE SNR-8000 is a portable receiver weighing around 4 kg, consumes 15-watt energy and is suitable for field use. It has 8 parallel channels on L1 and L2. It provides code and phase data on both frequencies and has a codeless option. Full P-code tracking provides highest precision phase and pseudo rages measurements, codeless tracking is automatic 'full back' mode. The code less mode uses the fact that each carrier has identical modulation of P-code/Y-code and hence the L1 signal can be cross-correlated with the L2 signal. Results are the differential phase measurement (L1-L2) and the group delay measurement (P1-P2).

5.7.5. Accuracy Specifications Are

P-Code pseudo range 1cm (5 minutes integration) Codeless pseudo range 10cm (5 minutes integration) Carrier phase 0.2 - 0.3 mm.

Codeless phase 0.2 - 0.7 mm.

One of the important features is that less than 1 cycle slip is expected for 100 satellite hours.

Navigation Receivers

Navigation receivers are rapidly picking up the market. In most cases a single C/A code sequencing or multiplexing channel is used. However, modules with four or five parallel channels are becoming increasingly popular. Position and velocity are derived from C/A code pseudoranges measurement and are displayed or downloaded to a personal computer. Usually neither raw data nor carrier phase information is available. Differential navigation is possible with some advanced models.

MAGELLAN NAV 1000 is a handheld GPS receiver and weighs only 850 grams. It was introduced in 1989 and later in 1990, NAV 1000 PRO model was launched. It is a single channel receiver and tracks 3 to 4 satellites with a 2.5 seconds update rate and has a RS 232 data port.

The follow up model in 1991 was NAV 5000 PRO. It is a 5-channel receiver tracking all visible satellites with a 1-second update rate. Differential navigation is possible. Carrier phase data can be used with an optional carrier phase module. The quadrifilar antenna is integrated to the receiver. Post processing of data is also possible using surveying receiver like ASHTECH XII located at a reference station. Relative accuracy is about 3 to 5 metres. This is in many cases sufficient for thematic purposes.

Many hand held navigation receivers are available with added features. The latest market situation can be obtained through journals like GPS world etc.

For most navigation purpose a single frequency C/A code receiver is sufficient. For accuracy requirements better than 50 to 100 meters, a differential option is essential. For requirement below 5 meters, the inclusion of carrier phase data is necessary. In high precision navigation the use of a pair of receivers with full geodetic capability is advisable. The main characteristics of multipurpose geodetic receiver are summarized in Table 4.

Table 4: Overview of Geodetic Dual-frequency GPS Satellite Receiver (1992)

Receiver	Channel		Code		Wavelen		Anti-spoofing
	L1	L2	L1	L2	L1	L2	
TI 4100	4	4	P	P			Single
MACROMET	6	6	-	-		/2	No influence
ASHTECH	12	12	C/A	-		/2	No influence
ASHTECH P	12	12	C/A,	P			Squaring
TRIMBLE	8-12	8-12	C/A	-		/2	No influence
TRIMBLE	9-12	9-12	C/A,	P			Codeless SSE
WM 102	7	1	C/A	P			Squaring
WILD GPS	9	9	C/A	p			Codeless
TURBO	8	8	C/A,	P			Codeless

Some of the important features for selecting a geodetic receiver are :

- Tracking of all satellites
- Both frequencies
- Full wavelength on L2
- Low phase noise-low code noise
- High sampling rate for L1 and L2
  - High memory capacity
- Low power consumption
- Full operational capability under anti spoofing condition

Further, it is recommended to use dual frequency receiver to minimize ion-spherical influences and take advantages in ambiguity solution.

5.8. GPS Surveying Differential Theory

Differential positioning is technique that allows overcoming the effects of environmental errors and SA on the GPS signals to produce a highly accurate position fix.

Accuracy

In general, an SPS receiver can provide position information with an error of less than 25 meter and velocity information with an error less than 5 meters per second. Upto 2 May 2000 U.S Government has activated Selective Availability (SA) to maintain optimum militar y effectiveness. Selective Availability inserts random errors into the ephemeris information broadcast by the satellites, which reduces the SPS accuracy to around 100 meters.

For many applications, 100-meter accuracy is more than acceptable. For applications that require much greater accuracy, the effects of SA and environmentally produced errors can be overcome by using a technique called Differential GPS (DGPS), which increases overall accuracy.

Differential Theory

Differential positioning is technique that allows overcoming the effects of environmental errors and SA on the GPS signals to produce a highly accurate position fix. This is done by determining the amount of the positioning error and applying it to position fixes that were computed from collected data.

Typically, the horizontal accuracy of a single position fix from a GPS receiver is 15 meter RMS (root-mean Square) or better. If the distribution of fixes about the true position is circular normal with zero mean, an accuracy of 15 meters RMS implies that about 63% of the fixes obtained during a session are within 15 meters of the true position.

Types of Errors

There are two types of positioning errors: correctable and non-correctable. Correctable errors are the errors that are essentially the same for two GPS receivers in the same area. Non-correctable errors cannot be correlated between two GPS receivers in the same area.

Table 5: Error Sources

Error Sources

Error Source	Approx. Equivalent Range Error (RMS) in meters
Correctable with Differential	
Clock (Space Segment)	3.0
Ephemeris (Control Segment)	2.7
Ionospheric Delay (Atmosphere)	8.2
Tropospheric Delay (Atmosphere)	1.8
Selective Availability (if implemented)	27.4
Total	28.9
Non-Correctable with Differential	
Receiver Noise (Unit)	9.1
Multipath (Environmental)	3.0
Total	9.6
Total user Equivalent range error (all sources)	30.5
Navigational Accuracy (HDOP = 1.5)	45.8

5.9. Differential GPS

Most DGPS techniques use a GPS receiver at a geodetic control site whose position is known. The receiver collects positioning information and calculates a position fix, which is then compared to the known co-ordinates. The difference between the known position and the acquired position of the control location is the positioning error.

Because the other GPS receivers in the area are assumed to be operating under similar conditions, it is assumed that the position fixes acquired by other receivers in the area (remote units) are subject to the same error, and that the correction computed for the control position should therefore be accurate for those receivers. The correction is communicated to the remote units by an operator at the control site with radio or cellular equipment. In post-processed differential, all units collect data for off-site processing; no corrections are determined in the field. The process of correcting the position error with differential mode is shown in the Figure.

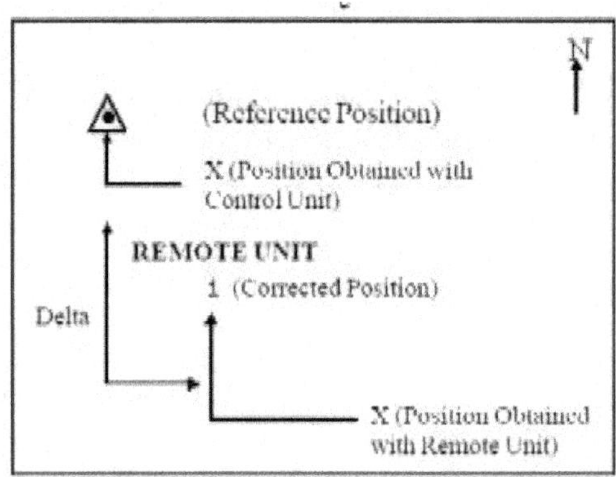

The difference between the known position and acquired position at the control point is the DELTA correction. DELTA, which is always expressed in meters, is parallel to the surface of the earth. When expressed in local co- ordinate system, DELTA uses North-South axis (y) and an East-West axis (x) in 2D operation; an additional vertical axis (z) that is perpendicular to the y and x is used in 3D operation for altitude.

5.9.1. Applications of GPS

- Providing Geodetic control.
- Survey control for Photogrammetric control surveys and mapping.
- Finding out location of offshore drilling.

- Pipeline and Power line survey.
- Navigation of civilian ships and planes. z Crustal movement studies.
- Geophysical positioning, mineral exploration and mining. z Determination of a precise geoid using GPS data.
- Estimating gravity anomalies using GPS.
- Offshore positioning: shiping, offshore platforms, fishing boats etc.

Astronomical observation of celestial bodies was one of the standard methods of obtaining coordinates of a position. This method is prone to visibility and weather condition and demands expert handling. Attempts have been made by USA since early 1960's to use space based artificial satellites. System TRANSIT was widely used for establishing a network of control points over large regions. Establishment of modern geocentric datum and its relation to local datum was successfully achieved through TRANSI T. Rapid improvements in higher frequently transmission and precise clock signals along with advanced stable satellite technology have been instrumental for the development of global positioning system.

The NAVSTAR GPS (Navigation System with Time and Ranging Global Positioning System) is a satellite based radio navigation system providing precise three- dimensional position, course and time information to suitably equipped user.

UNIT V

MODERN SURVEYING

MODEL SHORT QUESTIONS AND ANSWERS

1. **Compare between microwave and electro-optical EDM systems. (April/May 2018)**

Microwave EDM system	Electro-optical EDM system
- Microwave used as carrier wave	- Infra red beam used as carrier wave
- Used to measure distances Varying from 50m to 50km	- Measure lengths varying from 1km to 60km
- Comparatively less accurate	- More accurate system

2. **What is space segment? (April/May 2018)**

1. The space segments (SS) is composed of the orbiting GPS satellite, or space vehicles (SV).
2. This was modified to six planes with four satellite each.
3. The GPS design originally called for 24SVeight each in three circular orbital planes.

3. **What is multipath error? (April/May 2018)**

   1. A signal that bounces of a smooth object and hits the receiver antenna.

   2. It increases the length of time for a signal to reach the receiver.

4. **List the components of an electro-optical EDM system. (April/May 2018) (Nov/Dec 2014)**

   - Visible light produced by a tungsten lamp
   - Xeon flash tube or laser light or infra-red light
   - A light modulator
   - Optical parts for transmitting and receiving the modulated light
   - Photomultiplier
   - Phase meter
   - A read out unit

5. **Define total station and basic principles of Total Station. (April/May 2018) (Nov/Dec 2017) (April/May 2017)**

   The total station or electronic tacheometeris a combination of an electronic theodolite, an electronic distance measuring device and a microprocessor unit. With this device one can determine angles and distances from the instrument to the points to be surveyed. With the application of trigonometry the angles and distances may be used to compute the actual positions of surveyed points in absolute terms.

6. **What is mean by Selective Availability? (Nov/Dec2017)**

   Selective Availability (SA) was an intentional degradation of public GPS signals implemented for national security reasons.

7. **What is called anti spoofing? (Nov/Dec 2017) (Nov/Dec 2015) (April/May 2015)**

   - The process of S/A dither is adopted where by some of the satellite clocks are placed out of phase by amounts known only to the military. The process of S/A epsilon is applied so that the satellite returns an incorrect position for itself. These effects can reduce the accuracy of position fixing with the C/A code. In addition, since 1994, it has been made available to military users only. This is achieved by a process known as anti-spoofing in which the code is encrypted to a secret Y-code.

8. **Define GPS. (April/May 2017)**

   - Global positioning system is a space based all weather radio navigation system that provides quickly, accurately and in-expensively the time, position and velocity of the object anywhere on the globe at anytime.

9. **What do you understand from the satellite configuration? (April/May 2017) (Nov/Dec 2015)**

   - Depending on the model and configuration of your GSG unit, and the scenario chosen, several satellite systems can be simulated in a scenario, each of which you may want to configure in accordance with the requirements for your receiver-under-test.

10. **List out the various segments in GPS? (April/May 2015)**

    - Satellite constellation or space segment
    - Operational control segment
    - User equipment segment

11. **What sources of error in GPS? (Nov/Dec 2016)**

    - Satellite-related errors
    - Propagation-medium related errors
    - Receiver-related errors

12. **Write short notes on hand-held receivers. (April/May 2013)**

    - Handheld receiver is a data recorder which may be capable of tracing up to six satellites. The Leica MX 8600 series is a typical of this type of equipment. For differential work this must be used with the MX 8650 base station, which has the capability of tracking 12 satellites or any other compatible unit. Typical uses are for asset mapping by utilities and local authorities, where the base station would be permanently mounted at the office or depot

13. **What I sGPS?**

Global Positioning System (GPS) is a space-based all weather radio navigation system that provides quickly, accurately and in-expensively the time, position and velocity of the object anywhere on the globe at any time.

14. **List the advantage of GPS surveys.**

    - Three dimensional.
    - Site intervisibility not needed.

- Weather independent.
- Day or night operation.
- Common reference system.
- Rapid data processing with quality control.
- High precision.

15. Mention any three aims of GPS.

- To explain how GPS can be used to take code and place measurements to determine position and be able to explain the different between these.
- To identify the various sources of error in GPS and explain how each of these affects the accuracy obtained.
- To understand the reasons why differential and relative methods are essential for high precision surveying with GPS.

16. What is space segment?

Space segment consists of a group of earth-orbiting satellites. GPS uses a system of 24 satellites, arranged in six orbital planes, inclined at 55° to the equator at altitudes of about 20000km with orbital periods of about 12hours. Each satellite has a microprocessor on board for self-monitoring and data processing.

17. Explain user segment.

User segment is the total user and supplier community, both civilian and military. It consists of all earth-based GPD receivers, which vary greatly in size and complexity. The typical receiver comprises of an antenna and a pre amplifier, radio signal micro processor, control and display device, data recording unit and power supply.

18. Explain absolute positioning.

This type is based on single receiver station. As ranging is carried out strictly between the satellite and the receiver station it is also called as stand- alone type. Thus the positions are subjected to errors inherent in satellite positioning. The accuracy is in the range of 50 to100m.

19. What are the satellite-related errors?

GPS satellites are provided with accurate atomic clocks. Although they are accurate they may not be perfect. Because of slight inaccuracies in the time keeping may lead to inaccuracies in position measurements.

The satellite position in space is also important equally because it is stacking point of all of the GPS calculations. Although GPS satellites are placed into very high orbits, they are still drift slightly from their predicted orbits contributing to the errors.

20. Why GPS signal is so complicated?

- The signal is so complicated that it almost looks like random electrical noise and name it as "Pseudo-Random."
- Distance to a satellite is determined by measuring how long a radio signal takes to reach us from that satellite.
- To make the measurement we assume that both the satellite and our receiver are generating the same pseudo-random codes at exactly the same time.

21. Define Geo coding.

Geo coding" is the process of converting street addresses or other locations (ZIP codes, postal codes, city & state, airport IATA/ICAO codes, etc.) to latitude and longitude, which can be entered into a GPS device or geographical software. GPS Visualize offers several options for geo coding your information.

22. Write the principles of GPS.

GPS is a satellite based navigation system. It uses a digital signal at about 1.5 GHz from each satellite to send data to the receiver. The receiver can then deduce its exact range from the satellite, as well as the geographic position (GP) of the satellite. The GP is the location on the Earth directly below the satellite. This establishes a line of position (LOP) on the Earth.

23. Explain static technique of surveying with GPS.

Static technique of surveying was the first high precision method developed for GPS and is the standard GPS method for determining the length of baselines that are longer than 20km.

Using this method, the reference receiver is located at a known control point and a rover, that is set up at a point whose location is to be determined.

24. What are satellite-related errors?

- GPS satellite is provided with very accurate atomic clocks. Although they are accurate they may not be perfect. Because of slight inaccuracies in position measurements.
- The satellite position in space is also important equally because it is stacking point of all of the GPS calculations. Although GPS satellites are placed into very high orbits, they are still drift slightly from their predicted orbits contributing to the errors.

25. What is multipath error?

As the signal arrives at the surface of the earth it may get reflected by local obstructions and goes to the receiver's antenna through more than one path. This type of error is called multiple error as the signal is getting to the antenna by multiple paths.

26. What is meant by hand held GPS?

A handheld GPS is a device that uses the Global Positioning System, combining modern geographic technology with a portable, user-friendly device for everyday use. Features on some models may also provide information on geographic locations like national and historic landmarks. The device is often used by outdoors enthusiasts to pinpoint the coordinates of a certain location for future reference.

27. Define GPS navigation.

GPS, which stands for Global Positioning System, is a radio navigation system that allows land, sea, and airborne users to determine their exact location, velocity, and time 24 hours a day, in all weather conditions, anywhere in the world.

28. What are the sources of error in GPS?

- Satellite-related errors.
- Propagation-medium related errors.
- Receiver-related errors.
- Selective availability(S/A).

29. Mention the components of GPS?

- Satellite constellation called space segment.
- Ground control/monitoring network called operational control segment.
- User receiving equipment called user equipment segment.

30. Define Satellite.

An artificial body placed in orbit around the earth or moon or another planet in order to collect information or for communication. An object launched to orbit Earth or another celestial body, as a device for reflecting or relaying radio signals or for capturing images.

31. Explain Rapid-static Technique.

GPS is now the preferred method for control surveys on large construction sites. For these surveys, the static method previously described but the GPS receiver needs to be left for shorter occupation times of 10 to 30min.

32. Define Kinematic Technique.

It is used when a lot of points are to be surveyed in a relatively small area and where the accuracy required is not as high for static surveys. These include detail surveying (mapping) and construction measurements.

33. What is a Total Station?

A Total Station or an Electronic Tachometer is a combination of an Electronic Theodolite, an Electronic Distance Measuring Device (EDM) and a microprocessor with memory unit. With this device are can determine angles (both horizontal and vertical) and distance from the instrument to the points to be surveyed.

34. What is a carrier wave?

EDMs consist of a transmitter set up at one end of the lengths to be measured, sending out a continuous wave to the receiver at the other end. This wave, termed the carrier wave, is then modulated and the length determined.

35. List the components of an Electro-optical EDM system.

Main components of Electro-optical EDM instruments are:

- Visible light produced by a tungsten lamp, xenon flash tube or laser light or infra-red light.
- Optical parts for transmitting and receiving the modulated light.
- Photomultiplier.
- Phase meter and
- A read out unit.
- A light source.

36. What is called trilateration in Modern positioning system?

It is basically a technique of triangulation and no angular measurement are made. The three sides of the triangles are measured precisely using electronic distance measuring equipment or a total station. It is useful when measurement is difficult or impossible due to any reason.

37. What is a microwave system?

Microwave system held an important position in land surveying because they were used to measure distances varying from 50m to 50km. The introduction of GPS has diminished that importance.

38. Compare the microwave and the electro-optical systems adopted in total station.

Electro-optical instruments are more accurate than the microwave instruments, because the shorter the carrier wave length, the better is the accuracy of the total station. In bad atmospheric conditions and for long distance measurements, microwave instruments may be utilized for better penetration through fog and haze.

39. What are the salient parameters of a total station?

A total station comprises of an electronic theodolite with an EDM and a microprocessor. The theodolite measures the horizontal angle (Hz) and the vertical angle (V) of the line of sight from the centre of the total station to the centre of a target on point to measure. The intersection of the rotation of the axis of the horizontal and vertical circles in the centre of the station.

40. What is meant by EDM?

The total station or electronic tacheometer is a combination of an electronic theodolite, an electronic distance measuring device (EDM) and a microprocessor with memory unit. With this device one can determine angles and distances from the instrument to the points to be surveyed.

41. What are the software applications can be done by total station?

The software applications available on may total stations include the following:

- Slope correction and reduced orientation.
- Horizontal circle orientation.
- Co-ordinate measurement.
- Traverse measurements.
- Resection.
- Remote elevation measurement.

42. List out the sources of error in total station.

- Horizontal collimation or line of sight error.
- Tilting axis error.
- Compensator index error.
- Vertical collimation or vertical axis error.

43. List out the different types of Total stations.

- Manual total station.
- Semi-automatic total station.
- Automatic total station.

- Servo-driven total station.

44. Explain the parts of total station?

- Telescope.
- Plate level.
- Keyboard and display window.
- Tribrach.
- Vertical tangent screw.
- Power switch.
- Vertical Motion clamp.
- Horizontal screw.
- Horizontal motion clamp and in built EDM.

45. Explain the sources of error in total stations.

- Horizontal collimation or Line of sight error.
- Tilting axis error.
- Compensator index error.
- Vertical collimation or Vertical index error.
- Calibration procedure of total stations.
- General equipment maintenance.

46. List the disadvantages of a total station.

- Hard copies of field notes are not provided.
- During the process of surveyor to look over and check the work.
- For an overall check of the survey, it is to be done in the office using appropriate software.
- It cannot be used routinely for observations of the sun, unless special filter are used.

47. Explain the use of target prisms.

Prisms are used to reflect transmitted signals. A single reflector is a cube corner prism that has the characteristics of reflecting light rays precisely back to the existing instrument.

48. Explain basic principles of Total station.

Total station one can determine angles and distances from the instruments to the points to be surveyed. With the application of trigonometry the angles and distances may be used to compute the actual positions of surveyed points in absolute terms.

49. Mention the function of carrier frequency in three distinct EDM categories.

- Low-frequency radio systems.
- Microwave radio systems.
- Visible and infra-red light system.

50. Explain the pulse method adopted in EDM.

Pulse method is essentially based on the simple concept that the distance is a product of velocity and time. Here, a short but intense pulse of radiation id transmitted to a reflector target, which is transmitted back to the receiver along a parallel path. The measured distance is obtained from the product of velocity of the signal and the complete time taken fir the travel.

51. How horizontal angle is measured in a total station?

Horizontal angle is measured from the zero direction on the horizontal circle. Choice of the zero direction is generally made as instrument north. Instead one may set zero in the direction of the long axis of the map area or choose to orient the instrument approximately to true, magnetic or Grid North. Most total station can measure angle to at least 5 seconds or 0.0013888°.

52. Explain Tilting Axis error.

This axial error occurs when the tilting axis of the total station is not perpendicular to its vertical axis. This has no effect on sightings taken when the telescope is horizontal, but introduces errors into horizontal circle readings when the telescope is tilted, especially for steep sighting.

MODEL DETAIL QUESTIONS

1) (a) Explain the working principle of initial setting of total station. **(April/May 2018)**

   (b) Write short notes on traversing and Trilateration of total station.

2) Explain the principle underlying "Electronic Distance Measurement". Write a note on errors in EDM.

   (April/May 2018)(Nov/Dec 2017) (May/June 2014)

3) What are various sources of errors in GPS? Explain them. **(April/May 2018) (Nov/Dec 2014)**

4) Write short notes on types of GPS receivers. **(April/May 2018) (April/May 2017)**

5) Explain the various segments comprising the functioning of GPS with neat sketches. **(Nov/Dec 2017) (April/May 2017) (April/May 2015)**

6) With a suitable sketch, explain the salient features of Hand held and geodetic receivers. **(April/May 2017) (Nov/Dec 2016) (Nov/Dec 2015) (April/May 2015)**

7) Write short notes on task of control segment. **(April/May 2017)**

8) Explain in detail about care and maintenance of total station instruments. **(Nov/Dec 2016)**

9) Explain the working principle of geodimeter in total station. **(Nov/Dec 2017) (April/May 2015)**

10) Explain satellite configurations and signal structure with neat diagrams. **(Nov/Dec 2016)**

QUESTION PAPER CODE: 80068

B.E/B.TECH DEGREE EXAMINATION, APRIL/MAY 2019

THIRD SEMESTER

CIVIL ENGINEERING

CE 8351 – SURVEYING

PART A — (10 X 2 = 20 MARKS)

1. Define an "agonic and isogonic line".

2. Find the combined correction for curvature and refraction for a distance of a (a) 3400 m (b) 1.29 km.

3. What is the basic principle followed in stadia method?

4. What are the factors on which the choice of contour interval depends?

5. Differentiate between laplace station from satellite station.

6. Find the most probable value of the angle A from the following observation equation

$$A = 40°20'12''$$

$$2A = 80°40'20''$$

$$6A = 40°20'12''$$

7. Name the factors to be considered in the selection of a discharge site.

8. What is the relation between the right ascension and hour angle?

9. Bring out the temporary adjustment of a total station.

10. List the advantages of GPS surveying.

PART B — (5 x 13 = 65 MARKS)

11.(a)	(i) A chain line PQ intersects a pond. Two points A and B are taken on the chain line on opposite sides of the pond. A line AC, 250 m long, is set out on the left of AB and other line AD, 300 m long is set out on the right of AB. Points C, B and D are in the same straight line. CB and BD are 100 m and 150 m long respectively. Calculate the length of AB. (7)
	(ii) A traverse ABCDA is made in the form of a square taking in clockwise order. If the bearing of AB is 120° 30', find the bearing of the other sides. (6)

OR

(b)	A tacheometer is set up at an intermediate point on a traverse course AB and the following observations are taken on a staff held vertically. The instrument is fitted with an analytic lens and the multiplying constants is 100. The reduced level of A being given as 350.75 m. Calculate the length of AB and reduced level B. Staff reading Bearing Vertical Angle Intercept Axial hair reading A 40° 35' - 4° 24' 2.172 1.962 B 220° 35' -5° 12' 1.986 1.866 (13)

12.(a)	The top (Q) of a chimney was sighted {rom two stations P and R at very different levels, the stations P and R being in line with the top of the chimney. The angle of elevation from P to the top of the chimney was 38° 21' and that from R to the top of the chimney was 21° 18'. The angle of the elevation from R to a vane 2m above the foot of the staff held at P was 15° 11'. The heights of instrument at P and R were 1.87m and 1.64m respectively. The horizontal distance between P and R was 127 m and the reduced level of R was 112.78 m. Find the RL of the top of the chimney and the horizontal distance from P to the chimney. (13)

OR

b)	The following observations were made using a tacheometer fitted with an anallactic lens.

Instrument station	Height of instrumentation	Staff station	WCB	Vertical angle	Stadia Hair readings	Remarks
0	1.550	A	30°30'	4° 30'	1.155,	
					1.755,	RL of O
					2.355	= 150.
	1.550	B	75°	10° 15'	1.250,	000 m
			30'		2.000,	
					2.750	

Calculate the distance AB, RL of A and B. Find also the gradient of line AB. (13)

13.(a)	A satellite station S is 6.5 m from the main station A and the following observations were taken A = 0° 0'; B = 102°48'; C = 256°12'; D = 324°6'. The length AB, AC and AFM were computed to be 1895 m, 2277 m, 2522 m respectively. Determine the direction of the line AB, AC and AD. (13)
	OR
(b)	The following angles were measured at a station O so as to close the horizon: AOB = 83° 42' 28.75" weight 3 BOC = 102° 15' 43.26" weight 2 COD = 94° 38' 27.22" weight 4 DOA = 79° 23' 23.77" weight 2 Adjust the angle by method of correlates. (13)
14.(a)	Explain the different coordinates systems by which the position of heavenly body can be specified. (13)
	OR
(b)	Explain various sounding methods in detail. (13)
15.(a)	(i) Explain the working principle of a total station. (8) (ii) Discuss the different sources of errors in a total station. (5)
	OR
(b)	Explain in detail about the different segments of GPS. (13)
	PART C — (1 x 15 = 15 marks)
16.(a)	Discuss the field procedures involved in preparing a map of a proposed hospital building.
	OR
(b)	With a neat sketch, explain the types of surveying for the construction of highway include all possible measurements required.

B.E/B.Tech Degree Examination, November/December 2019

THIRD SEMESTER

CIVIL ENGINEERING

CE 8351 – SURVEYING

PART A — (10 X 2 = 20 MARKS)

1. Differentiate between true bearing and magnetic bearing

2. What is meant by balancing of sights?

3. What is the need for providing analytic lens?

4. Define contour interval.

5. What do you mean by reduction to centre?

6. Find the most probable value and the probable error of the area of a circle whose radius is 15.40 m = 0.02 m

7. What is celestial sphere?

8. List the different solution for a three point problem.

9. What is a total station?

10. What is the need for anti-spoofing in GPS?

PART B — (5 x 13 = 65 MARKS)

11.(a)	Describe the different equipment required for ranging and chaining and explain the different methods of ranging. (13)
	OR
(b)	Explain the temporary adjustment of a level. How is the reduction of levels and booking of staff readings done using the rise and fall system? (13)
12.(a)	A tacheometer is set up at an intermediate point on a transverse course AB and the following observations are taken on the staff held vertically. The instrument is fitted with an analytic lens and the multiplying constant is 100. The reduced level of A being given as 350.75 m. calculate the length of AB and the reduced level of B. Staff station Bearing Vertical angle intercept Axial hair reading A 40° 35' -4 ° 24' 2.172 1.962 B 220° 35' -5 ° 12' 1.986 1.866 (13)
	OR

(b)	To determine the elevation of the top of the aerial pole, the following observations were made: Inst station Reading on BM angle of elevation Remarks A 1.377 11 ° 53' RL of BM = 30.150 m B 1.263 8 °5' Station A and B and the top of the aerial pole are in the same vertical plane. Find the elevation of the top of the aerial pole, If the distance between A and B was 30 m. (13)
13.(a)	Find the most probable values of the following angles closing the horizontal at a station P = 45° 23' 37" weight 1 Q= 75° 37' 15" weight 2 R = 125° 21' 21" weight 3 S = 113° 37' 29" weight 3 (13)
	OR
(b)	In measuring angle at a triangulation C, it was found necessary to set the transit over another station P south west of C and 3 m from C, so that the angle APB is approximately bisected by the line PC. The angle APC and CPB were found to be 28° 20' 35" and 31° 26' 45" respectively. The side AB was computed to be 975 m in the adjacent triangle, and when the station C was observed the mean values of the angle CAB and CBA were recorded as 61° 30' 25" and 58° 34' 20" respectively. Determine the angle ABC (13)
14.(a)	What is meant by sounding? Describe briefly any four methods of locating soundings. (13)
	OR
(b)	Describe different types of celestial coordinate system (13)
15.(a)	With neat sketch explain the working model of a modern total station. (13)
	OR
(b)	Explain the various components of GPS and its working principles. (13).
	PART C —. (1 x 15 = 15 marks)
16.(a)	What are the sources of error in surveying? Explain the different precautions and correction procedures that can be adopted to eliminate the errors
	OR
(b)	List the characteristics of contour lines and the uses of contouring. Also explain the different methods of locating and interpolating contours.

1. Isogonic lines are the lines having the same

   A. elevation

   B. bearing

   C. declination

   D. dip

Ans: C

2. The variation of magnetic declination within a day is called

   A. Diurnal variation

   B. Irregular variation

   C. Annual variation

   D. Secular variation

Ans: A

3. Local attraction at a place may be due to

   A. Key bunches

   B. Steel buttons

   C. Current carrying bare wire

   D. Electric storm

Ans: C

4. The amount of correction due to local attraction at a place

   A. Is constant for all bearing

   B. Varies with bearing

   C. Changes from time to time

   D. Sometimes additive and sometimes subtractive

Ans: A

5. An instrumental error in compass surveying is because of

A. Radiation

B. Intersection

C. resection

D. traversing

Ans: B

6. line of collimation

A. Is the same as line of sight

B. The line joining the point of intersection of cross hairs and optical centre of object glass

C. The geometrical axis of the telescope

D. The line parallel to the bubble axis tube

Ans: B

7. The very first reading taken is called

A. Back sight

B. Fore sight

C. Intermediate sight

D. invert

Ans: A

8. A change point is

A. The very first station

B. The last station

C. The intermediate station where FS and BS are taken

D. The station after which the instrument is sifted

Ans: C

9. A leveling station is a place where

A. The level is set up

B. The level staff is held

C. Both FS and BS is taken

D. Temporary adjustments are done

Ans: B

10. The telescope of a dumpy level

A.   Is rigidly fixed to the leveling head

B.   Can be tilted in vertical plane

C.   Can be taken of its supports and reversed

D.   Permits interchange of eye piece and object glass

Ans: A

11. A bench mark is a

A.   Reference point

B.   The very first station

C.   The last station where the survey closes

D.   Point of known elevation

Ans: D

12. cross section and longitudinally sectioning is

A.   Simple levelling

B.   differential levelling

C.   Profile levelling

D.   Check levelling

Ans: B

13. Height of instrument method of booking reading is adopted

A.   When less number of intermediate sight exist

B.   In profile leveling

C.   In reciprocal leveling

D.   In differential leveling

Ans: C

14. An invert is taken when the point is

A.   Having high elevation

B.   Above the line of sights

C.   Below the line of sight

D.   Below ground level

Ans: A

15. In a survey it was recorded that $\sum$Rise = 0, then

A.   The ground is sloping

B.   It is continuously rising

C.   It is continuously falling

D.   The survey had many invert reading

Ans: B

16. When the staff is held on BM of RL 100.000, the staff reading was 2.000. When the staff is held on station P, the reading was 3.000. Hence HI is

A.   100.000

B.   102.000

C.   103.000

D.   99.000

Ans: C

17. An example of instrumental error in leveling is

A.   Earths curvature and atmospheric refraction

B.   Collimation error

C.   Wearing of shoe of level staff

D.   Defective tripod

Ans: B

18. when the temperature rises, length of bubble

A.   Remalns unaltered

B.   decereases

C.   increases

D.   Sometimes increase and sometimes decrease

Ans: B

19. reciprocal leveling eliminates

A.   Collimation error

B.   Collimation, curvature and refraction error

C.   curvature and refraction error

D.   Collimation and curvature error fully and refraction error partly

Ans: D

20. A contour map of the area is essential before proceeding with the construction of

A.   A building

B.   A swimming pool

C.   A dam

D.   A bridge

Ans: C

21. The slope correction for a length of 30 m along a gradient of 1 in 20, is

A.   3.75 cm

B.   0.375 cm

C.   37.5 cm

D.   2.75 cm

Ans: A

22. For the construction of highway (or railway)

A.   longitudinal sections are required

B.   cross sections are required

C.   both longitudinal and cross sections are required

D.   None of these

Ans: C

23. The branch of surveying in which both horizontal and vertical positions of a point, are determined by making instrumental observations, is known

A.   tacheometry

B.   tachemetry

C.   telemetry

D.   all the above.

Ans: D

24. In levelling operation

A.   when the instrument is being shifted, the staff must not be moved

B.   when the staff is being carried forward, the instrument must remain stationary

C.   Both (a) and (b)

D.   Neither (a) nor (b)

Ans: C

25. For setting out a simple curve, using two theodolites.

   A.   offsets from tangents are required

   B.   offsets from chord produced are required

   C.   offsets from long chord are required

   D.   deflection angles from Rankine's formula are required

   E.   none of these

Ans: E

26. Geodetic surveying is undertaken

   A.   for production of accurate maps of wide areas

   B.   for developing the science of geodesy

   C.   making use of most accurate instruments and methods of observation

   D.   for determination of accurate positions on the earth's surface of system of control points

   E.   all the above.

Ans: E

27. True meridian of different places

   A.   converge from the south pole to the north pole

   B.   Converge from the north pole to the south pole

   C.   converge from the equator to the poles

   D.   run parallel to each other.

Ans: C

28. The diaphragm of a stadia theodolite is fitted with two additional

   A.   horizontal hairs

   B.   vertical hairs

   C.   horizontal and two vertical hairs

   D.   none of these.

Ans: A

29. The best method of interpolation of contours, is by

A. estimation

B. graphical means

C. computation

D. all of these.

Ans: C

30. Setting out a curve by two theodolite method, involves

A. linear measurements only

B. angular measurements only

C. Both linear and angular measurements

D. none of these.

Ans: B

31. A relatively fixed point of known elevation above datum, is called

A. bench mark

B. datum point

C. reduced level

D. reference point.

Ans: A

32. Cross hairs in surveying telescopes, are fitted

A. in the objective glass

B. at the centre of the telescope

C. at the optical centre of the eye piece

D. in front of the eye piece.

Ans: D

33. Metric chains are generally available in

A. 10 m and 20 m length

B. 15 m and 20 m length

C. 20 m and 30 m length

D. 25 m and 100 m length

Ans: C

34. On a diagonal scale, it is possible to read up to

A.   one dimension

B.   two dimensions

C.   three dimensions

D.   Four dimensions

Ans: C

35. The main plate of a transit is divided into 1080 equal divisions. 60 divisions of the vernier coincide exactly with 59 divisions of the main plate. The transit can read angles accurate upto

A.   5"

B.   10"

C.   15"

D.   20"

Ans: D

36. A clinometer is used for

A.   measuring angle of slope

B.   correcting line of collimation

C.   setting out right angles

D.   defining natural features.

Ans: A

37. The construction of optical square is based, on the principle of optical

A. reflection

B. refraction

C. double refraction

D. double reflection.

Ans: D

38. Cross-staff is used for

A.   setting out right angles

B.   measuring contour gradient

C.   taking levels

D.   measuring distances

Ans: A

39. The minimum range for sliding the focusing lens in the internal focusing telescope for focusing at all distances beyond 4 m is

A.   5 mm

B.   10 mm

C.   15 mm

D.   20 mm

Ans: D

40. The slope correction may be ignored if

A.   the slope of the ground is less than 3°

B.   to slope of the ground is say 1 in 19

C.   both (a) and (b)

D.   neither (a) nor (b)

Ans: C

41. Which one of the following mistakes/errors may be cumulative + or - :

A.   bad ranging

B.   bad straightening

C.   erroneous length of chain

D.   sag.

Ans: C

42. The whole circle bearing of a line is 290°. Its reduced bearing is

A.   N 20° E

B.   N 20° W

C.   N 70° W

D.   S 70° E

Ans: C

43. The reduced bearing of a line is N 87° W. Its whole circle bearing is

A.   87°

B.   273°

C.   93°

D.   3°

Ans: B

44. *ABCD* is a rectangular plot of land. If the bearing of the side *AB* is 75°, the bearing of *DC* is

A. 75°

B. 255°

C. 155°

D. 285°

Ans: A

45. Grid lines are parallel to

A. magnetic meridian of the central point of the grid

B. line representing the central true meridian of the grid

C. geographical equator

D. none of these.

Ans: B

46. An angle of deflection right, may be directly obtained by setting the instrument to read

A. zero on back station

B. 180° on back station

C. 90°

D. 270° on back station.

Ans: A

47. The method of finding out the difference in elevation between two points for eliminating the effect of curvature and refraction is

A. reciprocal levelling

B. precise levelling

C. differential levelling

D. flying levelling

Ans: A

48. Perpendicularity of an offset may be judged by eye, if the length of the offset is

A. 5 m

B. 10 m

C. 15 m

D. 20 m

Ans: C

49. Ranging in chain survey means

  A.   looking at an isolated point not on the line

  B.   establishing an intermediate point on the line

  C.   determining the distance between end points

  D.   determining the offset distance

Ans: B

50. Greater accuracy in linear measurements, is obtained by

  A.   tacheometry

  B.   direct chaining

  C.   direct taping

  D.   all the above.

Ans: B

51. During chaining along a straight line, the leader of the party has 4 arrows in his hand while the follower has 6. Distance of the follower from the starting point is

  a) 4 chains
  b) 6 chains
  c) 120 m
  d) 180m

Ans: b

52. A metallic tape is made of

  a) steel
  b) invar
  c) linen
  d) cloth and wires

Ans: d

53. For a well-conditioned triangle, no angle should be less than

a) 20°
b) 30°
c) 45°
d) 60°

Ans: b

54. The angle of intersection of the two plane mirrors of an optical square is

a) 30°

b) 45°

c) 60°

d) 90°

Ans: b

55. The allowable length of an offset depends upon the

  a) degree of accuracy required

  b) method of setting out the perpendiculars and nature of ground

  c) scale of plotting

  d) all of the above

Ans: d

56. Which of the following angles can be set out with the help of French cross staff?

a)　45° only

b)　90° only

c)　either 45° or 90°

d)　any angle

Ans: c

57. Which of the following methods of offsets involves less measurement on the ground?

a)　method of perpendicular offsets

b)　method of oblique offsets

c)　method of ties

d)　all involve equal measurement on the ground

Ans: a

58. The point on the celestial sphere vertically below the observer's position, is called

a)　zenith

b)　celestial point

c)　nadir

d)　pole

Ans: c

59. The stereo plotting instruments are generally manufactured on the principle of

a) optical projection

b) optical mechanism projection

c) mechanical projection

d) All the above

Ans: d

60. The permissible error in chaining for measurement with chain on rough or hilly ground is

a) 1 in 100

b) 1 in 250

c) 1 in 500

d) 1 in 1000

Ans: b

61. The correction for sag is

a) always additive

b) always subtractive

c) always zero

d) sometimes additive and sometimes subtractive

Ans: b

62. Cross staff is an instrument used for

a) measuring approximate horizontal angles

b) setting out right angles

c) measuring bearings of the lines

d) none of the above

Ans: b

63. Normal tension is that pull which

a) is used at the time of standardising the tape

b) neutralizes the effect due to pull and sag

c) makes the correction due to sag equal to zero

d) makes the correction due to pull equal to zero

Ans: b

64. Which of the following is not used in measuring perpendicular offsets?

a) line ranger

b) steel tape

c) optical square

d) cross staff

Ans: a

65. If the length of a chain is found to be short on testing, it can be adjusted by

a) straightening the links

b) removing one or more small circular rings

c) closing the joints of the rings if opened out

d) all of the above

Ans: a

66. The maximum tolerance in a 20 m chain is

a) ±2 mm

b) ±3 mm

c) ±5 mm

d) ±8 mm

Ans: c

67. For accurate work, the steel band should always be used in preference to chain because the steel band

a) is lighter than chain

b) is easier to handle

c) is practically inextensible and is not liable to kinks when in use

d) can be easily repaired in the field

Ans: c

68. The length of a chain is measured from

a) centre of one handle to centre of other handle

b) outside of one handle to outside of other handle

c) outside of one handle to inside of other handle

d) inside of one handle to inside of other handle

Ans: b

69.   Select the incorrect statement.

a)   The true meridians at different places are parallel to each other.

b)   The true meridian at any place is not variable.

c)   The true meridians converge to a point in northern and southern hemispheres.

d)   The maps prepared by national survey departments of any country are based on true meridians.

Ans: a

70.   If the true bearing of a line AB is 269° 30′, then the azimuth of the line AB is a) 0° 30′

b) 89° 30′

c) 90° 30′

d) 269° 30′

Ans: c

71.   In the prismatic compass

 a) the magnetic needle moves with the box

 b) the line of the sight does not move with the box

 c) the magnetic needle and graduated circle do not move with the box

 d) the graduated circle is fixed to the box and the magnetic needle always remains in the N-S direction

Ans: c

72.   For a line AB

a)   the forebearing of AB and back bearing of AB differ by 180°

b)   the forebearing of AB and back bearing of BA differ by 180°

c)   both (a) and (b) are correct.

d)   none is correct

Ans: a

73.   Local attraction in compass surveying may exist due to

a)   incorrect levelling of the magnetic needle

b)   loss of magnetism of the needle

c)   friction of the needle at the pivot

d)   presence of magnetic substances near the instrument

Ans: d

74. In the quadrantal bearing system, a whole circle bearing of 293° 30′ can be expressed as

a) W23°30"N

b) N66°30"W

c) S113°30"N

d) N23°30"W

Ans: b

75. The prismatic compass and surveyor"s compass

a) give whole circle bearing (WCB) of a line and quadrantal bearing (QB) of a line respectively

b) both give QB of a line and WCB of a line

c) both give QB of a line

d) both give WCB of a line

Ans: a

76. The horizontal angle between the true meridian and magnetic meridian at a place is called

a) azimuth

b) declination

c) local attraction

d) magnetic bearing

Ans: b

77. A negative declination shows that the magnetic meridian is to the

a) eastern side of the true meridian

b) western side of the true meridian

c) southern side of the true meridian

d) none of the above

Ans: b

78. If the magnetic bearing of the sun at a place at noon in southern hemisphere is 167°, the magnetic declination at that place is

a) 77° N

b) 23° S

c) 13° E

d) 13° W

Ans: c

79.   The graduations in prismatic compass

i)   are inverted

ii)   are upright

iii)   run clockwise having 0° at south

iv)   run clockwise having 0° at north

The correct answer is

a)   (i) and (iii)

b)   (i) and (iv)

c)   (ii) and (iii)

d)   (ii) and (iv)

Ans: a

80.   Agate cap is fitted with a

a)   cross staff

b)   level

c)   chain

d)   prismatic compass

Ans: d

81.   The temporary adjustments of a prismatic compass are

i)   Centering

ii)   Levelling

iii)   Focusing the prism The correct order is

a) (0, (iii), 00

b) (0, (ii), (iii)

c) (ii), (iii), 0)

d) (in), (i), (ii)

Ans: b

82.   Theodolite is an instrument used for

a)   tightening the capstan-headed nuts of level tube

b)   measurement of horizontal angles only

c)   measurement of vertical angles only

d)   measurement of both horizontal and vertical angles

Ans: d

83. The process of turning the telescope about the vertical axis in horizontal plane is known
as
a) transiting
b) reversing
c) plunging
d) swinging

Ans: d

84. Size of a theodolite is specified by
a) the length of telescope
b) the diameter of vertical circle
c) the diameter of lower plate
d) the diameter of upper plate

Ans: c

85. Which of the following is not the function of levelling head ?
a) to support the main part of the instrument
b) to attach the theodolite to the tripod
c) to provide a means for leveling the theodolite
d) none of the above

Ans: d

86. If the lower clamp screw is tightened and upper clamp screw is loosened, the theodolite
may be rotated
a) on its outer spindle with a relative motion between the vernier and graduated scale of
lower plate
b) on its outer spindle without a relative motion between the vernier and gra-duated scale of
lower plate
c) on its inner spindle with a relative motion between the vernier and the graduated scale of
lower plate
d) on its inner spindle without a relative motion between the vernier and the graduated scale
of lower plate

Ans: c

87. A telescope is said to be inverted if its

a) vertical circle is to its right and the bubble of the telescope is down

b) vertical circle is to its right and the bubble of the telescope is up

c) vertical circle is to its left and the bubble of the telescope is down

d) vertical circle is to its left and the bubble of the telescope is up

Ans: a

88. The cross hairs in the surveying telescope are placed

a) midway between eye piece and objective lens

b) much closer to the eye-piece than to the objective lens

c) much closer to the objective lens than to the eye piece

d) anywhere between eye-piece and objective lens

Ans: b

89. For which of the following permanent adjustments of theodolite, the spire test is used?

a) adjustment of plate levels

b) adjustment of line of sight

c) adjustment of horizontal axis

d) adjustment of altitude bubble and vertical index frame

Ans: c

90. The adjustment of horizontal cross hair is required particularly when the instrument is used for

a) leveling

b) prolonging a straight line

c) measurement of horizontal angles

d) all of the above

Ans: a

91. Which of the following errors is not eliminated by the method of repetition of horizontal angle measurement?

a) error due to eccentricity of verniers

b) error due to displacement of station signals

c) error due to wrong adjustment of line of collimation and trunnion axis

d) error due to inaccurate graduation

Ans: b

92. The error due to eccentricity of inner and outer axes can be eliminated by

a) reading both verniers and taking the mean of the two

b) taking both face observations and taking the mean of the two

c) double sighting

d) taking mean of several readings distributed over different portions of the graduated circle

Ans: a

93. In the double application of principle of reversion, the apparent error is

a) equal to true error

b) half the true error

c) two times the true error

d) four times the true error

Ans: d

94. Which of the following errors can be eliminated by taking mean of bot face observations?

a) error due to imperfect graduations

b) error due to eccentricity of verniers

c) error due to imperfect adjustment of plate levels

d) error due to line of collimation not being perpendicular to horizontal axis

Ans: d

95. Which of the following errors cannot be eliminated by taking both face observations?

a) error due to horizontal axis not being perpendicular to the vertical axis

b) index error i.e. error due to imperfect adjustment of the vertical circle vernier

c) error due to non-parallelism of the axis of telescope level and line of collimation

d) none of the above

Ans: d

96. If a tripod settles in the interval that elapses between taking a back sight reading and the following foresight reading, then the elevation of turning point will

a) increase

b) decrease

c) not change

d) either „a" or „b"

Ans: a

97. If altitude bubble is provided both on index frame as well as on telescope of a theodolite, then the instrument is levelled with reference to

i)   altitude bubble on index frame

ii)   altitude bubble on index frame if it is to be used as a level

iii)   altitude bubble on telescope

iv)   altitude bubble on telescope if it is to be used as a level The correct answer is

a)   only (i)

b)   both (i) and (iv)

c)   only (iii)

d)   both (ii) and (iii)

Ans: b

98.   A"level line"is a

a)   horizontal line

b)   line parallel to the mean spheriodal surface of earth

c)   line passing through the center of cross hairs and the center of eye piece

d)   line passing through the objective lens and the eye-piece of a dumpy or tiltinglevel

Ans: b

99.   The following sights are taken on a "turning point"

a)   foresight only

b)   backsight only

c)   foresight and backsight

d)   foresight and intermediate sight

Ans: c

100. The rise and fall method of levelling provides a complete check on

a)   backsight

b)   intermediate sight

c)   foresight

d)   all of the above

Ans: d

101. If the R.L. of a B.M. is 100.00 m, the back- sight is 1.215 m and the foresight is 1.870 m, the R.L. of the forward station is

a) 99.345 m

b) 100.345 m

c) 100.655m

d) 101.870m

Ans: a

102. In an internal focussing type of telescope, the lens provided is

a) concave

b) convex

c) plano-convex

d) plano-concave

Ans: a

103. Which of the following errors can be neutralised by setting the level midway between the two stations?

a) error due to curvature only

b) error due to refraction only

c) error due to both curvature and re-fraction

d) none of the above

Ans: c

104. Height of instrument method of levelling is

a) more accurate than rise and fall method

b) less accurate than rise and fall method

c) quicker and less tedious for large number of intermediate sights

d) none of the above

Ans: c

105. The rise and fall method

a) is less. accurate than height of instrument method

b) is not suitable for levelling with tilting levels

c) provides a check on the reduction of intermediate point levels

d) quicker and less tedious for large number of intermediate sights

Ans: c

106. If the staff is not held vertical at a levelling station, the reduced level calculated from the observation would be

a) true R.L.

b) more than true R.L.

c) less than true R.L.

d) none of the above

Ans: c

107. The difference between a level line and a horizontal line is that

a) level line is a curved line while horizontal line is a straight line

b) level line is normal to plumb line while horizontal line may not be normal to plumb line at the tangent point to level line

c) horizontal line is normal to plumb line while level line may not be normal to the plumb line

d) both are same

Ans: a

108. The sensitivity of a bubble tube can be increased by

a) increasing the diameter of the tube

b) decreasing the length of bubble

c) increasing the viscosity of liquid

d) decreasing the radius of curvature of tube

Ans: a

109. With the rise of temperature, the sensitivity of a bubble tube

a) decreases

b) increases

c) remains unaffected

d) none of the above

Ans: a

110. Refraction correction

a) completely eliminates curvature correction

b) partially eliminates curvature correction

c) adds to the curvature correction

d) has no effect on curvature correction

Ans: b

111. The R.L, of the point A which is on the floor is 100 m and back sight reading on A is 2.455 m. If the foresight reading on the point B which is on the ceiling is 2.745 m, the R.L. of point B will be

a) 94.80 m

b) 99.71 m

c) 100.29 m

d) 105.20 m

Ans: d

112. As applied to staff readings, the corrections for curvature and refraction are respectively, The above table shows a part of a level field book. The value of X should be

a) 98.70

b) 100.00

c) 102.30

d) 103.30

Ans: b

113. If the horizontal distance between the staff point and the point of observation is d, then the error due to curvature of earth is proportional to

a) d

b) 1/d

c) d2

d) 1/d2

Ans: c

114. Sensitiveness of a level tube is designated by

a) radius of level tube

b) length of level tube

c) length of bubble of level tube

d) none of the above

Ans: a

115. Which of the following statements is in-correct?

a) Error due to refraction may not be completely eliminated by reciprocal levelling.

b) Tilting levels are commonly used for precision work.

c) The last reading of levelling is always a foresight.

d) All of the above statements are incorrect.

Ans: d

116. Dumpy level is most suitable when

a) the instrument is to be shifted frequently

b) fly levelling is being done over long distance

c) many readings are to be taken from a single setting of the instrument

d) all of the above

Ans: c

117. The difference of levels between two stations A and B is to be determined. For best results, the instrument station should be

a) equidistant from A and B

b) closer to the higher station

c) closer to the lower station

d) as far as possible from the line AB

Ans: a

118. Contour interval is

a) inversely proportional to the scale of the map

b) directly proportional to the flatness of ground

c) larger for accurate works

d) larger if the time available is more

Ans: a

119. An imaginary line lying throughout the surface of ground and preserving a constant inclination to the horizontal is known as

a) contour line

b) horizontal equivalent

c) contour interval

d) contour gradient

Ans: d

120. The suitable contour interval for a map with scale 1 : 10000 is

a) 2 m

b) 5m

c) 10 m

d) 20 m

Ans: a

121. Select the correct statement.

a) A contour is not necessarily a closed curve.

b) A contour represents a ridge line if the concave side of lower value contour lies towards the higher value contour.

c) Two contours of different elevations do not cross each other except in case of an overhanging cliff.

d) All of the above statements are correct.

Ans: c

122. A series of closely spaced contour lines represents a

a) steep slope

b) gentle slope

c) uniform slope

d) plane surface

Ans: a

123. Direct method of contouring is

a) a quick method

b) adopted for large surveys only

c) most accurate method

d) suitable for hilly terrains

Ans: c

124. In direct method of contouring, the process of locating or identifying points lying on a contour is called

a) ranging

b) centring

c) horizontal control

d) vertical control

Ans: d

125. In the cross-section method of indirect contouring, the spacing of cross-sections depends upon

i) contour interval

ii) scale of plan

iii) characteristics of ground The correct answer is

a) only (i)

b) (i)and(ii)

c) (ii) and (iii)

d) (i), (ii) and (iii)

Ans: d

126. Which of the following methods of con-touring is most suitable for a hilly terrain ?

a) direct method

b) square method

c) cross-sections method

d) tacheometric method

Ans: d

127. Select the correct statement.

a) Contour interval on any map is kept constant.

b) Direct method of contouring is cheap¬er than indirect method.

c) Inter-visibility of points on a contour map cannot be ascertained.

d) Slope of a hill cannot be determined with the help of contours.

Ans: a

128. Closed contours, with higher value inwards, represent a

a) depression

b) hillock

c) plain surface

d) none of the above

Ans: b

129. Contour interval is

a) the vertical distance between two consecutive contours

b) the horizontal distance between two consecutive contours

c) the vertical distance between two points on same contour

d) the horizontal distance between two points on same contour

Ans: a

130. Benchmark is established by

a) hypsometry

b) barometric levelling

c) spirit levelling

d) trigonometrical levelling

Ans: c

131. The type of surveying which requires least office work is

a) tacheomefry

b) trigonometrical levelling

c) plane table surveying

d) theodolite surveying

Ans: c

132. Intersection method of detailed plotting is most suitable for

a) forcsts

b) urban areas

c) hilly areas

d) plains

Ans: c

133. Detailed plotting is generally done by

a) radiation

b) traversing

c) resection

d) all of the above

Ans: a

134. Three point problem can be solved by

a) Tracing paper method

b) Bessels method

c) Lehman"s method

d) all of the above

Ans: d

135. The size of a plane table is

a) 750 mm x 900 mm

b) 600 mm x 750 mm

c) 450 mm x 600 mm

d) 300 mm x 450 mm

Ans: b

136. The process of determining the locations of the instrument station by drawing re sectors from the locations of the known stations is called

a) radiation

b) intersection

c) resection

d) traversing

Ans: c

137. The instrument used for accurate centering in plane table survey is

a) spirit level

b) alidade

c) plumbing fork

d) trough compass

Ans: c

138. Which of the following methods of plane table surveying is used to locate the position of an inaccessible point?

a) radiation

b) intersection

c) traversing

d) resection

Ans: b

139. The two point problem and three point problem are methods of

a) resection
b) orientation
c) traversing
d) resection and orientation

Ans: d

140. The resection by two point problem as compared to three point problem

a) gives more accurate problem
b) takes less time
c) requires more labour
d) none of the above

Ans: c

141. The methods used for locating the plane table stations are

i) radiation
ii) traversing
iii) intersection
iv) resection

The correct answer is

a) (i) and (ii)
b) (iii) and (iv)
c) (ii) and (iv)
d) (i) and (iii)

Ans: c

142. After fixing the plane table to the tripod, the main operations which are needed at each plane table station are

i) levelling
ii) orientation
iii) centering

The correct sequence of these operations is a) (i), (ii),.(iii)

b) (i), (iii), (ii)
c) (iii), (i), (ii)
d) (ii), (Hi), (i)

Ans: b

143. Bowditch rule is applied to

a)   an open traverse for graphical adjustment

b)   a closed traverse for adjustment of closing error

c)   determine the effect of local attraction

d)   none of the above

Ans: b

144. If in a closed traverse, the sum of the north latitudes is more than the sum of the south latitudes and also the sum of west departures is more than the sum of the east departures, the bearing of the closing line is in the

a)   NE quadrant

b)   SE quadrant

c)   NW quadrant

d)   SW quadrant

Ans: b

145. If the reduced bearing of a line AB is N60°W and length is 100 m, then the latitude and departure respectively of the line AB will be

a) +50 m, +86.6 m

b) +86.6 m, -50 m

c) +50m, -86.6 m

d) +70.7 m,-50 m

Ans: b

146. The angle between the prolongation of the preceding line and the forward line of a traverse is called

a)   deflection angle

b)   included angle

c)   direct angle

d)   none of the above

Ans: a

147. Transit rule of adjusting the consecutive coordinates of a traverse is used where

a) linear and angular measurements of the traverse are of equal accuracy

b) angular measurements are more accurate than linear measurements

c) linear measurements are more accurate than angular measurements

d) all of the above

Ans: b

148. Which of the following methods of theodolite traversing is suitable for locating the details which are far away from transit stations ?

a) measuring angle and distance from one transit station

b) measuring angles to the point from at least two stations

c) measuring angle at one station and distance from other

d) measuring distance from two points on traverse line

Ans: b

149. Subtense bar is an instrument used for

a) levelling

b) measurement of horizontal distances in plane areas

c) measurement of horizontal distances in undulated areas

d) measurement of angles

Ans: c

150. Horizontal distances obtained by thermometric observations

a) require slope correction

b) require tension correction

c) require slope and tension corrections

d) do not require slope and tension corrections

Ans: d

151. The number of horizontal cross wires in a stadia diaphragm is

a) one

b) two

c) three

d) four

Ans: c

152. If the intercept on a vertical staff is ob-served as 0.75 m from a tacheometer, the horizontal distance between tacheometer and staff station is

a) 7.5 m

b) 25 m

c) 50

d) 75 m

Ans: d

153. For a tacheometer the additive and multi-plying constants are respectively

a) 0 and 100

b) 100 and 0

c) 0 and 0

d) 100 and 100

Ans: a

154. If the focal length of the object glass is 25 cm and the distance from object glass to the trunnion axis is 15 cm, the additive constant is

a) 0.1

b) 0.4

c) 0.6

d) 1.33

Ans: b

155. Overturning of vehicles on a curve can be avoided by using

a) compound curve

b) vertical curve

c) reverse curve

d) transition curve

Ans: d

156. Different grades are joined together by a

a) compound curve

b) transition curve

c) reverse curve

d) vertical curve

Ans: d

157. To avoid large centering error with very short legs, observations are generally made

(A) To chain pins

(B) By using optical system for centering the theodolite

(C) To a target fixed on theodolite tripod on which theodolite may be fitted easily

(D) All the above

Ans: Option C

158. Different grades are joined together by a

(A) Compound curve

(B) Transition curve

(C) Reverse curve

(D) Vertical curve

Ans: Option D

159. Which of the following methods of theodolite traversing is suitable for locating the details which are far away from transit stations?

(A) Measuring angle and distance from one transit station

(B) Measuring angles to the point from at least two stations

(C) Measuring angle at one station and distance from other

(D) Measuring distance from two points on traverse line

Ans: Option B

160. Which of the following methods of contouring is most suitable for a hilly terrain?

(A) Direct method

(B) Square method

(C) Cross-sections method

(D) Tachometric method

Ans: Option D

161. The chord of a curve less than peg interval, is known as

(A) Small chord

(B) Sub-chord

(C) Normal chord

(D) Short chord

Ans: Option B

162. The size of a plane table is

(A) 750 mm × 900 mm

(B) 600 mm × 750 mm

(C) 450 mm × 600 mm

(D) 300 mm × 450 mm

Ans: Option B

163. If the reduced bearing of a line AB is N60°W and length is 100 m, then the latitude and departure respectively of the line AB will be

(A) +50 m, +86.6 m

(B) +86.6 m, -50 m

(C) +50 m, -86.6 m

(D) +70.7 m, -50 m

Ans: Option B

164. Contour interval is

(A) Inversely proportional to the scale of the map

(B) Directly proportional to the flatness of ground

(C) Larger for accurate works

(D) Larger if the time available is more

Ans: Option A

165. The smaller horizontal angle between the true meridian and a survey line, is known

(A) Declination

(B) Bearing

(C) Azimuth

(D) Dip

Ans: Option C

166. Contour interval is

(A) The vertical distance between two consecutive contours

(B) The horizontal distance between two consecutive contours

(C) The vertical distance between two points on same contour

(D) The horizontal distance between two points on same contour

Ans: Option A

167. Check lines (or proof lines) in Chain Surveying, are essentially required

(A) To plot the chain lines

(B) To plot the offsets

(C) To indicate the accuracy of the survey work

(D) To increase the out-turn

Ans: Option C

168. After fixing the plane table to the tripod, the main operations which are needed at each plane table station are (i) Levelling (ii) Orientation (iii) Centering The correct sequence of these operations is

(A) (i), (ii), (iii)

(B) (i), (iii), (ii)

(C) (iii), (i), (ii)

(D) (ii), (iii), (i)

Ans: Option B

169. There are two stations A and B. Which of the following statements is correct?

(A) The fore bearing of AB is AB

(B) The back bearing of AB is BA

(C) The fore and back bearings of AB differ by 180°

(D) All the above

Ans: Option D

170. The Random errors tend to accumulate proportionally to

(A) Numbers of operations involved

(B) Reciprocal of operations involved

(C) Square root of the number of operation involved

(D) Cube root of the number of operation involved

Ans: Option C

171. In direct method of contouring, the process of locating or identifying points lying on a contour is called

(A) Ranging

(B) Centring

(C) Horizontal control

(D) Vertical control

Ans: Option D

172. Pick up the incorrect statement from the following:

(A) While measuring a distance with a tape of length 100.005 m, the distance to be increasing by 0.005 m for each tape length

(B) An increase in temperature causes a tape to increase in length and the measured distance is too large

(C) The straight distance between end points of a suspended tape is reduced by an amount called the sag correction

(D) A 100 m tape of cross section 10 mm × 0.25 mm stretches about 10 mm under 5 kg pull

Ans: Option B

173. For the construction of highway (or railway)

(A) Longitudinal sections are required

(B) Cross sections are required

(C) Both longitudinal and cross sections are required

(D) None of these

Ans: Option C

174. Detailed plotting is generally done by

(A) Radiation

(B) Traversing

(C) Resection

(D) All of the above

Ans: Option A

175. If the smallest division of a vernier is longer than the smallest division of its primary scale, the vernier is known as

(A) Direct vernier

(B) Double vernier

(C) Retrograde vernier

(D) Simple vernier

Ans: Option C

176. The method of reversal

(A) Is usually directed to examine whether a certain part is truly parallel or perpendicular to another

(B) Makes the erroneous relationship between parts evident

(C) Both (a) and (b)

(D) Neither (a) nor (b)

Ans: Option C

177. The line normal to the plumb line is known as

(A) Horizontal line

(B) Level line

(C) Datum line

(D) Vertical line

Ans: Option B

178. In levelling operation

(A) When the instrument is being shifted, the staff must not be moved

(B) When the staff is being carried forward, the instrument must remain stationary

(C) Both (a) and (b)

(D) Neither (a) nor (b)

Ans: Option C

179. The two point problem and three point problem are methods of

(A) Resection

(B) Orientation

(C) Traversing

(D) Resection and orientation

Ans: Option D

180. Ramsden eye-piece consists of

(A) Two convex lenses short distance apart

(B) Two concave lenses short distance apart

(C) One convex lens and one concave lens short distance apart

(D) Two Plano-convex lenses short distance apart, with the convex surfaces facing each other

Ans: Option D

181. The line of sight is kept as high above ground surface as possible to minimise the error in the observed angles due to

(A) Shimmering

(B) Horizontal refraction

(C) Vertical refraction

(D) Both shimmering and horizontal refraction

Ans: Option D

182. If is the stadia distance, is the focal length and is the distance between the objective and vertical axis of the techeometer, the multiplying constant, is

(A) f/i

(B) i/f

(C) $(f + d)$

(D) f/d

Ans: Option A

183. The desired sensitivity of a bubble tube with 2 mm divisions is 30". The radius of the bubble tube should be

(A) 13.75 m

(B) 3.44 m

(C) 1375 m

(D) None of these

Ans: Option A

184. Subtense bar is an instrument used for

(A) Levelling

(B) Measurement of horizontal distances in plane areas

(C) Measurement of horizontal distances in undulated areas

(D) Measurement of angles

Ans: Option C

185. Volume of the earth work may be calculated by

(A) Mean areas

(B) End areas

(C) Trapezoidal

(D) All the above

Ans: Option D

186. Pick up the correct statement from the following:

(A) The directions of plumb lines suspended at different points in a survey are not strictly parallel

(B) In surveys of small extent, the effect of curvature may be ignored and the level surface of the earth is assumed as horizontal

(C) In surveys of large extent, the effect of curvature of the earth must be considered

(D) All the above

Ans: Option D

187. In levelling operation

(A) If second reading is more than first, it represents a rise

(B) If first reading is more than second, it represents a rise

(C) If first reading is less than second, it represents a fall

(D) Both (b) and (c)

Ans: Option D

188. Over-turning of vehicles on a curve can be avoided by using

(A) Compound curve

(B) Vertical curve

(C) Reverse curve

(D) Transition curve

Ans: Option D

189. Which of the following introduces an error of about 1 in 1000 if 20 m chain is used?

(A) Length of chain 20 mm wrong

(B) One end of the chain 0.9 m off the line

(C) One end of chain 0.9 m higher than the other

(D) All the above

Ans: Option D

190. Chain surveying is well adopted for

(A) Small areas in open ground

(B) Small areas with crowded details

(C) Large areas with simple details

(D) Large areas with difficult details

Ans: Option A

191. Two contour lines, having the same elevation

(A) Cannot cross each other

(B) Can cross each other

(C) Cannot unite together

(D) Can unite together

Ans: Option D

192. Which of the following statements is incorrect?

(A) Error due to refraction may not be completely eliminated by reciprocal levelling

(B) Tilting levels are commonly used for precision work

(C) The last reading of levelling is always a foresight

(D) All of the above statements are incorrect

Ans: Option D

193. In levelling operation,

(A) The first sight on any change point is a back sight

(B) The second sight on any change point is a fore sight

(C) The line commences with a fore sight and closes with a back sight

(D) The line commences with a back sight and closes with a foresight

Ans: Option D

194. The length of a traverse leg may be obtained by multiplying the latitude and

(A) Secant of its reduced bearing

(B) Sine of its reduced bearing

(C) Cosine of its reduced bearing

(D) Tangent of its reduced bearing

Ans: Option A

195. While working on a plane table, the correct rule is:

(A) Draw continuous lines from all instrument stations

(B) Draw short rays sufficient to contain the points sought

(C) Intersection should be obtained by actually drawing second rays

(D) Take maximum number of sights as possible from each station to distant objects

Ans: Option B

196. The vertical angle between longitudinal axis of a freely suspended magnetic needle and a horizontal line at its pivot, is known

(A) Declination

(B) Azimuth

(C) Dip

(D) Bearing

Ans: Option C

197. In the cross-section method of indirect contouring, the spacing of cross-sections depends upon (i) Contour interval (ii) Scale of plan (iii) Characteristics of ground The correct answer is

(A) Only (i)

(B) (i) and (ii)

(C) (ii) and (iii)

(D) (i), (ii) and (iii)

Ans: Option D

198. Transition curves are introduced at either end of a circular curve, to obtain

(A) Gradually decrease of curvature from zero at the tangent point to the specified quantity at the junction of the transition curve with main curve

(B) Gradual increase of super-elevation from zero at the tangent point to the specified amount at the junction of the transition curve with main curve

(C) Gradual change of gradient from zero at the tangent point to the specified amount at the junction of the transition curve with main curve

(D) None of these

Ans: Option B

199. Tilt of the staff in stadia tacheometry increases the intercept if it is

(A) Away from the telescope pointing down hill

(B) Towards the telescope pointing up-hill

(C) Away from the telescope pointing up-hill

(D) None of these

Ans: Option C

200. In a lemniscate curve the ratio of the angle between the tangent at the end of the polar ray and the straight, and the angle between the polar ray and the straight is

(A) 2

(B) 3

(C) 4/3

(D) 3/2

Ans: Option D

201. A level when set up 25 m from peg A and 50 m from peg B reads 2.847 on a staff held on A and 3.462 on a staff held on B, keeping bubble at its centre while reading. If the reduced levels of A and B are 283.665 m and 284.295 m respectively, the collimation error per 100 m is

(A) 0.015 m

(B) 0.030 m

(C) 0.045 m

(D) 0.060 m

Ans: Option D

202. Hydrographic surveys deal with the mapping of

(A) Large water bodies

(B) Heavenly bodies

(C) Mountainous region

(D) Canal system

Ans: Option A

203. Bowditch rule is applied to

(A) An open traverse for graphical adjustment

(B) A closed traverse for adjustment of closing error

(C) Determine the effect of local attraction

(D) None of the above

Ans: Option B

204. Angles to a given pivot station observed from a number of traverse stations when plotted, the lines to the pivot station intersect at a common point

(A) Angular measurements are correct and not the linear measurements

(B) Linear measurements are correct and not the angular measurements

(C) Angular and linear measurements are correct and not the plotting of traverse

(D) Angular and linear measurements and also plotting of the traverse are correct

Ans: Option D

205. Pick up the correct statement from the following:

(A) The horizontal angle between magnetic meridian and true meridian at a place is called magnetic declination or variance of the compass

(B) The imaginary lines which pass through points at which the magnetic declinations are equal at a given time are called isogonic lines

(C) The isogonic lines through places at which the declination is zero are termedagonic lines

(D) All the above

Ans: Option D

206. Straight, parallel and widely spaced contours represent

(A) A steep surface

(B) A flat surface

(C) An inclined plane surface

(D) Curved surface

Ans: Option C

207. The real image of an object formed by the objective must lie

(A) In the plane of cross hairs

(B) At the centre of the telescope

(C) At the optical centre of the eye-piece

(D) Anywhere inside the telescope

Ans: Option A

208. For a tachometer the additive and multiplying constants are respectively

(A) 0 and 100

(B) 100 and 0

(C) 0 and 0

(D) 100 and 100

Ans: Option A

209. Correct distance obtained by an erroneous chain is:

(A) (Erroneous chain length/Correct chain length) × Observed distance

(B) (Correct chain length/Erroneous chain length) × Observed distance

(C) (Correct chain length/Observed distance) × Erroneous chain length

(D) None of these

Ans: Option A

210. An imaginary line lying throughout on the surface of the earth and preserving a constant inclination to the horizontal, is called

(A) Contour line

(B) Contour gradient

(C) Level line

(D) Line of gentle scope

Ans: Option B

211. The constant vertical distance between two adjacent contours, is called

(A) Horizontal interval

(B) Horizontal equivalent

(C) Vertical equivalent

(D) Contour interval

Ans: Option D

212. Question No. 65 If a 30 m chain diverges through a perpendicular distance d from its correct alignment, the error in length, is

(A) $(d^2/60)$ m (B) $(d^2/30)$ m (C) $(d^2/40)$ m (D) $(d/30)$ m

Ans: Option A

213. The sensitivity of a bubble tube can be increased by

(A) Increasing the diameter of the tube

(B) Decreasing the length of bubble

(C) Increasing the viscosity of liquid

(D) Decreasing the radius of curvature of tube

Ans: Option A

214. Pick up the correct statement from the following:

(A) It is difficult to eliminate an error completely at first trial

(B) Instability of the instrument makes it almost impossible to adjust it satisfactorily

(C) Adjustment screws must be left bearing firmly but should never be forced

(D) All the above

Ans: Option D

215. The properties of autogenous curve for automobiles are given by

(A) True spiral

(B) Cubic parabola

(C) Bernoulli's Lemniscate

 (D) Clothoid spiral

Ans: Option C

216. In chain surveying field work is limited to

(A) Linear measurements only

(B) Angular measurements only

(C) Both linear and angular measurements

(D) All the above

Ans: Option A

217. The difference of levels between two stations A and B is to be determined. For best results, the instrument station should be

(A) Equidistant from A and B

(B) Closer to the higher station

(C) Closer to the lower station

(D) As far as possible from the line AB

Ans: Option A

218. The curve composed of two arcs of different radii having their centres on the opposite side of the curve, is known

(A) A simple curve

(B) A compound curve

(C) A reverse curve

(D) A vertical curve

Ans: Option C

219. Pick up the correct statement from the following:

(A) An observation or the resulting reading with the level on a levelling staff is called sight

(B) A back sight is the first sight taken after setting up the instrument in any position

(C) The first sight on each change point is a fore sight

(D) All the above

Ans: Option D

220. The angle of intersection of a curve is the angle between

(A) Back tangent and forward tangent

(B) Prolongation of back tangent and forward tangent

(C) Forward tangent and long chord

(D) Back tangent and long chord

Ans: Option A

221. In tacheometrical observations, vertical staff holding is generally preferred to normal staffing, due to

(A) Ease of reduction of observations

(B) Facility of holding

(C) Minimum effect of careless holding on the result

(D) None of these

Ans: Option C

222. Whole circle bearing of a line is preferred to a quadrantal bearing merely because

(A) Bearing is not completely specified by an angle

(B) Bearing is completely specified by an angle

(C) Sign of the correction of magnetic declination is different in different quadrants

(D) Its Trigonometrical values may be extracted from ordinary tables easily

Ans: Option B

223. It is more difficult to obtain good results while measuring horizontal distance by stepping

(A) Up-hill

(B) Down-hill

(C) In low undulations

(D) In plane areas

Ans: Option A

224. Pick up the correct statement from the following:

(A) The apparent error on reversal is twice the actual error

(B) The correction may be made equal to half the observed discrepancy

(C) The good results may be obtained from a defective instrument by reversing and taking the mean of two erroneous results

(D) All the above

Ans: Option D

225. Which of the following methods of plane table surveying is used to locate the position of an inaccessible point?

(A) Radiation (B) Intersection (C) Traversing (D) Resection

Ans: Option B

226. Imaginary line passing through points having equal magnetic declination is termed as

(A) Isogon (B) Agonic line (C) Isoclinic line (D) None of these

Ans: Option A

227. In geodetic surveys higher accuracy is achieved, if

(A) Curvature of the earth surface is ignored

(B) Curvature of the earth surface is taken into account

(C) Angles between the curved lines are treated as plane angles

(D) None of these

Ans: Option B

228. Transit rule of adjusting the consecutive coordinates of a traverse is used where

(A) Linear and angular measurements of the traverse are of equal accuracy

(B) Angular measurements are more accurate than linear measurements

(C) Linear measurements are more accurate than angular measurements

(D) All of the above

Ans: Option B

229. Pick up the correct statement from the following:

(A) The lines of sight while observing back sight and fore sight lie in the same horizontal plane

(B) The staff readings are measurements made vertically downwards from a horizontal plane

(C) The horizontal plane with reference to which staff readings are taken, coincides with the level surface through the telescope axis

(D) All the above

Ans: Option D

230. The operation of resection involves the following steps

1. Rough orientation of the plane table

2. The three lines form a triangle of error

3. Drawing lines back through the three control points

4. Select a point in the triangle of error such that each ray is equally rotated either clockwise or anti clockwise

5. The points obtained by three rays are the correct location. The correct sequence is

(A) 1, 3, 2, 4, 5

(B) 1, 2, 3, 4, 5

(C) 1, 4, 3, 2, 5

(D) 1, 3, 2, 4, 5

Ans: Option A

231. For true difference in elevations between two points A and B , the level must be set up

(A) At any point between A and B

(B) At the exact midpoint of A and B

(C) Near the point A

(D) Near the point B

Ans: Option B

232. Rankine's deflection angle in minutes is obtained by multiplying the length of the chord by (A) Degree of the curve

(B) Square of the degree of the curve

(C) Inverse of the degree of the curve

(D) None of these

Ans: Option A

233. In case of a double line river, contours are

(A) Stopped at the banks of the river

(B) Stopped at the edge of the river

(C) Drawn across the water

(D) Drawn by parabolic curves having their vertex at the centre of the water

Ans: Option B

234. The branch of surveying in which both horizontal and vertical positions of a point, are determined by making instrumental observations, is known

(A) Tacheometry

(B) Tachemetry

(C) Telemetry

(D) All the above

Ans: Option D

235. Refraction correction

(A) Completely eliminates curvature correction

(B) Partially eliminates curvature correction

(C) Adds to the curvature correction

(D) Has no effect on curvature correction

Ans: Option B

236. If the plane table is not horizontal in a direction at right angles to the alidade, the line of sight is parallel to the fiducial edge only for

(A) Horizontal sights

(B) Inclined sights upward

(C) Inclined sight downward

(D) None of these

Ans: Option A

237. In optical reading instruments

(A) The vertical circle is usually continuous from $0°$ to $359°$

(B) The readings increase when the telescope is elevated in the face left position

(C) The readings decrease when the telescope is elevated in the face right position

(D) All the above

Ans: Option D

238. If the whole circle bearing of a line is 270°, its reduced bearing is

(A) N 90° W

(B) S 90° W

(C) W 90°

(D) 90° W

Ans: Option C

239. The process of determining the locations of the instrument station by drawing re sectors from the locations of the known stations is

(A) Radiation

(B) Intersection

(C) Resection

(D) Traversing

Ans: Option C

240. Surveys which are carried out to depict mountains, rivers, water bodies, wooded areas and other cultural details, are known as

(A) Cadastral surveys

(B) City surveys

(C) Topographical surveys

(D) Guide map surveys

Ans: Option C

241. The operation of revolving a plane table about its vertical axis so that all lines on the sheet become parallel to corresponding lines on the ground, is known

(A) Levelling

(B) Centering

(C) Orientation

(D) Setting

Ans: Option C

242. The number of horizontal cross wires in a stadia diaphragm is

(A) One

(B) Two

(C) Three

(D) Four

Ans: Option C

243. Total latitude of a point is positive if it lies

(A) North of the reference parallel

(B) South of the reference parallel

(C) East of the reference parallel

(D) West of the reference parallel

Ans: Option A

244. Under ordinary conditions, the precision of a theodolite traverse is affected by

(A) Systematic angular errors

(B) Accidental linear errors

(C) Systematic linear errors

(D) Accidental angular errors

Ans: Option C

245. The sum of the interior angles of a geometrical figure laid on the surface of the earth differs from that of the corresponding plane figure only to the extent of one second for every

(A) 100 sq. km of area

(B) 150 sq. km of area

(C) 200 sq. km of area

(D) None of these

Ans: Option C

246. An angle of deflection right, may be directly obtained by setting the instrument to read

(A) Zero on back station

(B) 180° on back station

(C) 90°

(D) 270° on back station

Ans: Option A

247. Location of contour gradient for a high way is best set out from

(A) Ridge down the hill

(B) Saddle down the hill

(C) Bottom to the ridge

(D) Bottom to the saddle

Ans: Option B

248. Setting out a curve by two theodolite method, involves

(A) Linear measurements only

(B) Angular measurements only

(C) Both linear and angular measurements

(D) None of these

Ans: Option B

249. Direct method of contouring is

(A) A quick method

(B) Adopted for large surveys only

(C) Most accurate method

(D) Suitable for hilly terrains

Ans: Option C

250. A lemniscate curve will not be transitional throughout, if its deflection angle, is
(A) 45° (B) 60° (C) 90° (D) 120°

Ans: Option A

www.ingramcontent.com/pod-product-compliance
Lightning Source LLC
LaVergne TN
LVHW051139200726
843495LV00022B/1731